METRIC IN MINUTES

The Comprehensive Resource for Learning the Metric System (SI)

Dennis R. Brownridge

Metric in Minutes

Copyright © 1994 by Professional Publications, Inc. All rights are reserved. No part of this publication may be reproduced, stored in a retrieval system, or transmitted, in any form or by any means, electronic, mechanical, photocopying, recording, or otherwise, without the prior written permission of the publisher.

Printed in the United States of America

ISBN: 0-912045-71-X

Professional Publications, Inc.
1250 Fifth Avenue, Belmont, CA 94002
(415) 593-9119

Current printing of this edition: 2

Library of Congress Cataloging-in-Publication Data

Brownridge, Dennis R., 1944–
    Metric in minutes : the comprehensive resource for learning
the metric system (SI) / Dennis R. Brownridge.
       p.   cm.
    Includes bibliographical references and index.
    ISBN 0-912045-71-X
    1. Metric system.   2. Metric system--Study and teaching.
  I. Title.
QC91.B955   1994
530.8'12--dc20   93-47071
                                      CIP

When you can measure what you are speaking about, and express it in numbers, you know something about it; but when you cannot measure it, when you cannot express it in numbers, your knowledge is of a meager and unsatisfactory kind; it may be the beginning of knowledge, but you have scarcely, in your thoughts, advanced to the stage of science.

—Lord Kelvin (1824–1907)

Contents

Part I The International System

Chapter 1 Introduction   *1*

1.1   SI: The Modern Metric System
1.2   Advantages of SI
1.3   Coherent Systems
1.4   Non-SI Units
1.5   SI in the United States
1.6   Converting Non-SI Units

Chapter 2 SI Structure and Mathematical Conventions   *4*

2.1   Quantity and Unit Symbols
2.2   Base Units
2.3   Supplementary Units
2.4   Derived Units
2.5   Digit Separators
2.6   Fraction Bars
2.7   Scientific Calculators

Chapter 3 Precision and Rounding   *8*

3.1   Significant Digits
3.2   Rounding Rules
3.3   Accuracy
3.4   Estimating Precision
3.5   Counting Numbers

Chapter 4 Prefixes   *11*

4.1   Advantages of Prefixes
4.2   Powers of Ten
4.3   Three Notations
4.4   Prefix Names and Symbols
4.5   Choosing Prefixes
4.6   Changing Prefixes by Hand
4.7   Changing Scientific Notation to a Prefix, by Hand
4.8   Handling Prefixes on Calculators
4.9   Prefixes That Are Not Multiples of 1000

Part II Quantities and Their Units

Chapter 5 Length   *17*

5.1   The Meter
5.2   Multiples of Meter

Chapter 6 Area   *21*

6.1   The Square Meter
6.2   Prefixes Hecto and Centi
6.3   Multiples of Square Meter
6.4   Changing Prefixes

Chapter 7 Volume   *25*

7.1   The Cubic Meter
7.2   Prefixes Hecto, Deka, Deci, and Centi
7.3   Alternative Names
7.4   Multiples of Cubic Meter
7.5   Changing Prefixes

Chapter 8 Mass and Density   *30*

8.1   The Kilogram
8.2   Prefixes and Alternative Names
8.3   Density
8.4   Buoyancy
8.5   Mixtures and Concentrations
8.6   Summary of Prefix Complications

Chapter 9 Time and Rates   *35*

9.1   The Second
9.2   Non-SI Time Units
9.3   Defining the Second
9.4   Time of Day
9.5   Speed and Velocity
9.6   Flow (Volumetric)

Chapter 10 Force   *38*

10.1   The Newton
10.2   Gravitation
10.3   Electromagnetic Force

10.4   Strong and Weak Forces
10.5   Torque

Chapter 11   Energy   *43*
11.1   The Joule
11.2   Mass-Energy Equivalence
11.3   Kinetic Energy

Chapter 12   Power   *46*
12.1   The Watt
12.2   Energy Efficiency

Chapter 13   Pressure and Stress   *49*
13.1   The Pascal
13.2   Gage versus Absolute

Chapter 14   Temperature and Heat   *52*
14.1   Heat
14.2   Temperature: The Kelvin
14.3   Degrees Celsius
14.4   Kelvin versus Degree Celsius
14.5   Heat Flow
14.6   Thermal Conductance
14.7   Thermal Insulance (Insulation Value)
14.8   Specific Heat
14.9   Latent Heat
14.10   Temperature in Ideal Gases
14.11   Defining the Kelvin

Chapter 15   Frequency   *58*
15.1   The Hertz
15.2   Electromagnetic Radiation

Chapter 16   Angles   *62*
16.1   Supplementary Units
16.2   The Radian
16.3   The Steradian

16.4   Degrees and Related Units (Non-SI)
16.5   Grads (Non-SI)
16.6   Gradients (Slopes)

Chapter 17   Electromagnetic Units   *65*
17.1   Charge: the Coulomb
17.2   Electric Current: the Ampere
17.3   Electric Potential: the Volt
17.4   Electric Resistance: the Ohm
17.5   Other Electromagnetic Units

Chapter 18   The Mole   *69*
18.1   Amount of Substance: the Mole
18.2   Atomic Mass Units (Non-SI)
18.3   Defining the Mole
18.4   Using Moles

Chapter 19   Physiological Units   *72*
19.1   Light
19.2   Ionizing Radiation

Chapter 20   Compound Units   *75*

**Appendix A   Rules for the Correct
                Use of SI   *77***

**Appendix B   Conversion Factors for
                Non-SI Units   *80***
Simple Conversions
Factor Label Method
Non-SI Systems
Alphabetical List of Non-SI Units

Answers to Problems   *92*

Bibliography   *99*

About the Author   *100*

Index   *101*

Tables

Inside Front Cover

1 Prefixes

2 Common Units

3 Alternative Names

Inside Back Cover

4 Specialized SI Units

5 Non-SI Units Approved for Use with SI

4.1 Exponent Rules

4.2 Prefixes and Their Equivalents

5.1 Multiples of the Meter

6.1 Common Multiples of the Square Meter

7.1 Multiples of Cubic Meter

8.1 Common Multiples of Kilogram

B.1 Some Non-SI Unit Systems

Figures

5.1 Original definition of the meter.

5.2 Length comparisons.

6.1 Multiples of the square meter make nesting squares.

6.2 Area comparisons.

7.1 Multiples of cubic meter make nesting cubes.

8.1 Mass comparisons.

10.1 Force, weight, and mass.

10.2 Torque.

11.1 Energy comparisons.

12.1 Power comparisons.

14.1 Kelvin and Celsius temperature scales.

15.1 Electromagnetic radiation.

16.1 Radian.

16.2 Steradian.

16.3 Degrees.

16.4 Grads.

16.5 Gradient.

Preface

This book grew out of my frustration from the lack of an up-to-date, authoritative, and comprehensive, yet clear and practical guide to the simplified, modern version of the metric system known as SI. It covers the whole system but is written for a general audience and entails no mathematics beyond simple high-school algebra. A background in science or technology is not required.

Part I gives an overview of SI, explains its logic and structure, and demonstrates simple methods for manipulating SI prefixes, both by hand and with a calculator. Part II describes individual quantities (types of measurement) and their units. They are discussed in logical order, building from the simplest and most common to the more complex and specialized.

Real-world examples illustrate the use of SI. Problem sets at the end of each chapter provide self-tests, with answers at the back of the book.

Appendix A summarizes the rules for correct use, including correct examples and common mistakes. Appendix B lists conversion factors for some five hundred non-SI units, with instructions for converting.

I am indebted to countless students on whom this material has been tested and refined. Their bewilderment over our traditional units and their enthusiasm for SI led me to write the book. I also wish to thank Marilyn Uffner, my editor at PPI, for her encouragement, perseverance, and unflagging good humor.

PART I

The International System

CHAPTER 1

Introduction

1.1 SI: THE MODERN METRIC SYSTEM

This book describes the basic physical quantities of nature and their measurement in the simplified, modern version of the metric system, called the *International System of Units*. The International System is used in science and by nearly everyone outside the United States. Its abbreviation is SI (without periods) from the initials of its French name, *Système International d'Unités*. SI was established in 1960 to replace the various metric systems, which had accumulated many redundant and noncoherent units. Although SI is popularly called "the metric system," many metric units are *not* SI. Avoid obsolete, non-SI metric units, many of which still appear—incorrectly—in textbooks, instruments, and product labels.

The complete International System is displayed on the front and back inside covers. The most common units are in Table 2; those limited to specialized fields are in Table 4. Alternative names for some unit multiples are in Table 3. Common compound units (generic units without special names) are listed in Chapter 20.

SI is governed by the General Conference of Weights and Measures (CGPM*), an international organization founded a century ago. The United States is a charter member. The CGPM has developed strict rules for using SI so that it will not degenerate into the chaos of our older units. These rules are described at appropriate places in the book and summarized in Appendix A, with correct examples and common mistakes.

1.2 ADVANTAGES OF SI

SI has been adopted throughout the world because of its great simplicity:

- It has only one unit for each quantity (type of measurement), so unit conversions are never necessary.
- Decimals are used exclusively, eliminating awkward fractions.

*From the initials of its French name, *Conférence Générale des Poids et Mesures.*

- Long, cumbersome rows of zeroes are replaced with simple letter prefixes.
- Units have simple, unambiguous letter symbols, which can be manipulated algebraically.
- It is an almost completely *coherent* system, which means that units are defined by the laws of nature and fit together in the simplest possible way.

Although SI is not a perfect system, it is the only comprehensive, practical, and coherent measurement scheme in use. It greatly simplifies math, science, technology, education, and everyday life. British educators report that adoption of SI has saved their students many class hours each year.

1.3 COHERENT SYSTEMS

A coherent measuring system has the smallest possible number of independent units. All the other units are defined as simple products or quotients of these base units, using the same equations or laws that define the quantities. For example, in a coherent system based on the inch, the cubic inch is the unit of volume (not the gallon, pint, or cubic foot). In a coherent system based on the foot, the square foot is the unit of area (whereas the acre, square yard, and square mile are not). Coherence makes it possible to manipulate units with simple algebra and to discover relationships that would be difficult to grasp otherwise. (For examples, see particularly Sections 14.5 through 14.9).

1.4 NON-SI UNITS

All units that do not belong to the International System are called *non-SI units*. Americans use hundreds of such units, most of which are not part of any real system. How much is an acre, an acre foot, or a board-foot? How does a troy ounce differ from an avoirdupois ounce or a fluid ounce? What is the difference between a long ton, short ton, register ton, freight ton, refrigeration ton, and nuclear ton? These illogical and confusing non-SI units are just a fraction of those used daily in the United States. To understand them all you would have to memorize thousands of conversion factors and complex definitions.

The non-SI units traditionally used in the United States are correctly called *inch-pound* units. The popular term "English system" is a misnomer, since most are not of English origin and are certainly not systematic. Moreover, the gallons, pints, quarts, fluid ounces, and tons used in the United States are very different from the Imperial units of the same name formerly used in all the other English-speaking nations (Britain, Canada, Australia, New Zealand, Ireland, etc.). All the other English-speaking countries have adopted SI and are phasing out non-SI units of any kind.

Indeed, one might argue that SI is more "English" than our traditional units, many of which were invented by the French, Romans, or Babylonians. Half of those SI units named after persons honor English or American scientists (the newton, watt, joule, kelvin, henry, gray, tesla, farad, and siemens).

1.5 SI IN THE UNITED STATES

The United States is the only industrialized nation in the world that does not require SI for general use. The U.S. Constitution (Article I, Section 8) explicitly gives Congress the power to establish measuring units, but no system has ever been enforced. In 1866 Congress legalized the metric system (now SI), which remains the country's only legal measuring *system*. However, innumerable individual non-SI units are legal as well. Ironically, all of our traditional units have been based on metric units since 1893. For example, an inch is defined as exactly 2.54 centimeters. In 1959 U.S. length units (inch, foot, yard, mile, etc.) were even shortened slightly to make them simpler multiples of the meter.

Other English-speaking countries adopted SI in the 1960s and 70s. The United States prepared to do so as well with the 1975 Metric Conversion Act (15 U.S.C. § 205b). A few major industries converted, but many people resisted. Some feared that conversion would be too difficult or expensive, although the experience of other nations has shown otherwise. For example, Canada was able to change every highway speed limit sign in the country over a single weekend, using decals. When India converted, it was found that even illiterate street vendors could master everything they needed to know in a few hours. Others opposed SI for irrational reasons, denouncing it as "un-American" or "One-Worldism." Under pressure from various groups, Congress again made SI voluntary, rendering the 1975 Act ineffective.

In the 1980s it became increasingly apparent that our chaotic, nonstandard, and noncoherent units were eroding the ability of American students, workers, and industries to compete in a global marketplace. Progressive business and education leaders urged the government to mandate SI once and for all. In the 1988 Omnibus Trade and Competitiveness Act (Pub. L. No. 100-418, § 5164) Congress established SI as the preferred system for U.S. trade and commerce and required all Federal agencies to adopt it by the end of 1992 (or as quickly as possible without undue hardship). SI remained voluntary for private business.

As the use of SI inevitably increases, it is essential that Americans learn how to use it correctly. Otherwise, its full benefits will not be realized and we will be faced with another reeducation task in the future.

1.6 CONVERTING NON-SI UNITS

There are no numerical conversions in pure SI because it has only one unit for each quantity. However, it is sometimes necessary to convert non-SI data into SI or vice versa. Appendix B describes the process and lists conversion factors for some five hundred non-SI units. A few of them, used worldwide for centuries, have been approved for permanent use with SI; they are listed in Table 5 on the inside back cover. Included are the natural time units, day and year, and five ancient Babylonian units: hours and minutes of time, and degrees, minutes, and seconds of angle (see Chapters 9 and 16).

Problem Set 1
Introduction

1. Give the full English name of SI.

2. Why is it poor practice to call SI "the metric system?"

3. Give five advantages of SI (five reasons it is easy to use).

4. What is the term used to describe a measurement system in which units are defined as simple products or quotients of a few fundamental base units?

5. What is the correct term for all units that are not part of the International System?

6. What is the correct term for the non-SI, non-metric units traditionally used in the United States (feet, pounds, gallons, etc.)?

7. What has been the only legal measuring *system* in the United States (since 1866)?

8. How are nearly all non-SI units now defined?

9. Why is it unnecessary to convert units within SI?

CHAPTER 2

SI Structure and Mathematical Conventions

2.1 QUANTITY AND UNIT SYMBOLS

A *quantity* is a physical attribute or phenomenon that can be measured, such as length, volume, mass, or time. A *unit* is a standard of measurement, such as a meter, kilogram, or second. Each unit has a *symbol* (m, kg, or s, for example). (The common quantities and their units are displayed on the inside covers of this book.)

A quantity is expressed as a number times a unit. For example, in SI, the quantity "length" is measured in meters (the symbol is m). Thus the expression "5 m" is a quantity consisting of the number "5" times the unit "meter." When writing this quantity, no multiplication symbol is used, but a space is always left between the number and the unit for legibility (write 5 m, not 5m). The unit must always accompany the number; a quantity without a unit is usually meaningless.

Never use fractions with SI units (write 5.5 m, not $5\frac{1}{2}$ m). Use the correct symbols, never abbreviations such as "sec" for second (s) or "cc" for cubic centimeter (cm^3). Do not add an "s" to symbols for the plural (write 10 kg, not 10 kgs). However, normal English plurals are used when spelling out units (25 kilograms, 15 meters, etc.). Do not use the plural if the number is less than 1 (write 0.5 meter, not 0.5 meters). Do not use a period after a symbol, except, of course, at the end of a sentence. Use a zero in front of a leading decimal point, so the point will not be overlooked (write 0.5 kg, not .5 kg). Units named after a person, such as the watt, newton, and hertz, have capitalized symbols—and so does the liter (L), to avoid confusion with the numeral 1. All other unit symbols are always lowercase, even when the surrounding text is capitalized. Spelled out units are common nouns and are not capitalized, even if their symbol *is* capitalized (write 15 newtons, not 15 Newtons). SI symbols are always written in upright (normal, or roman) letters, even if the surrounding text is italicized (slanted to the right).

Symbols are also used for quantities. For example, we may use *l* as the symbol for length in a formula or an equation, such as *l* = 5 m. Quantity symbols are written in slanted (oblique, or italic) letters. It is important to distinguish the two letter styles because you may have both quantity and unit symbols in the same problem. For example, a lowercase roman "m" is the symbol for meter (a unit) but a lowercase italic "*m*" is the symbol for mass (a quantity). Both occur when applying such common equations as Newton's Second Law of Motion:

$$F = ma = 50 \text{ kg} \cdot 9.8 \text{ m/s}^2$$

SI recognizes three classes of units—base, supplementary, and derived—as explained in the next three sections.

2.2 BASE UNITS

SI is built on seven fundamental, independent standards called *base units*:

Symbol	Base unit	Quantity
m	meter	length
kg	kilogram	mass
s	second	time
K	kelvin	temperature
A	ampere	electric current
cd	candela	luminous intensity
mol	mole	amount of substance

The SI base units are all arbitrary inventions. There is nothing special about them, nor are they inherently superior to any other units. They were adopted from old metric systems because it would have been impractical to change every measurement in the world. If we were developing SI from scratch today we would quite possibly invent different base units—and fewer of them. The base units are discussed in detail in Part II.

2.3 SUPPLEMENTARY UNITS

The two SI angle units, radian and steradian, are classed as *supplementary units*. They are considered supplementary because angles are not physical quantities like length or mass, but are instead ratios, or fractions. See Chapter 16 for a detailed discussion of the supplementary units.

2.4 DERIVED UNITS

All other SI units are derived as products or quotients of the base (or supplementary) units, using the same equation that defines the quantity. No numerical factors are involved.* For

*Two exceptions are the sievert (see Section 19.2) and the Celsius temperature scale (Section 14.3).

example, the quantity called *speed* is defined as length per time. The SI unit of speed is therefore a meter per second—the unit of length divided by the unit of time. ("Per" means "divided by.") In symbols, you can show division of units three different ways—with a horizontal bar, a slash, or a negative exponent. For example, 25 meters per second may be written as

$$\frac{25 \text{ m}}{\text{s}} \qquad 25 \text{ m/s} \qquad 25 \text{ m·s}^{-1}$$

The slash and negative exponent styles are convenient for writing the unit on one line, whereas the horizontal bar is usually more convenient when working problems.

To multiply unit symbols, use a raised dot (·). For example, the SI unit for length times temperature, meter kelvin, is written in symbols as m·K. The dot must be included when writing unit symbols because a symbol may consist of more than one letter. Otherwise, it would be impossible to tell which letters are part of the symbol and which are being multiplied:

$$\text{kg·m} \qquad \text{Pa·s} \qquad \text{mol·K}$$

However, the dot is usually omitted with quantity symbols (*italic letters*) because they have only one full-sized character (although small subscript characters may be attached). Multiplication is understood.

$$ab \qquad m_1 m_2 \qquad F_G d$$

In all cases, multiplication is neither pronounced nor spelled out (say "meter kelvin," not "meter times kelvin"). The final unit is pluralized, when necessary. For example, 15 Pa·s is pronounced and spelled out "15 pascal seconds." Write the multiplication dot well above the baseline of the letters and numbers and make it large enough that it will not be mistaken for a decimal point.

Do not use the old-fashioned multiplication sign, ×, in units. It is easily confused with the letter x and has been largely replaced by the dot in modern math texts for all applications. In this book, "×" is used only when referring to the multiplication key on a calculator.

Do not mix symbols with spelled-out names in the same expression. For example, you may write 5 kilograms per second *or* 5 kg/s (but *not* 5 kg/second *or* 5 kg per s).

Some derived units are used so often that, for convenience, they have been given special short names and symbols, such as newton (N) and watt (W). These units are listed with the base units in Tables 2 and 4. Most were named after scientists who did pioneering work on the quantity that the unit measures (e.g., Isaac Newton and James Watt). These units are discussed in detail in Part II, in the chapters on their respective quantities.

Derived units without special names are called *compound units*—meter per second, for example. An unlimited number of compound units can be formed by multiplying or dividing base or derived units in different ways. Common examples are listed in Chapter 20.

2.5 DIGIT SEPARATORS

The preferred international practice is to group digits in long numbers with spaces, not commas. Leave a space every three places both left *and* right of the decimal point. On computers, type "hard spaces," which will not break and wrap to the next line or expand when full-justifying.

$$12\,000\,000 \qquad 6.783\,217$$

The space is usually omitted if there are *only* four digits left or right of the decimal.

$$5300 \qquad 3.6452$$

This practice is recommended by the U.S. government media guide (*American National Standard for Metric Practice;* see Bibliography) and followed in most public-school science textbooks. Most countries use a comma for the decimal point, so inserting commas as digit separators can lead to great confusion if the data is used internationally.

2.6 FRACTION BARS

In many of the examples in this book, we will set up problems in "fraction bar" notation. This style keeps a complex problem neatly organized and reduces clutter when there are many compound units and powers of ten. It simplifies solution with a calculator and makes it easy to cancel units, thereby providing a check that the problem has been set up correctly. In fraction bar style, factors are separated with vertical bars instead of pairs of parentheses or dots. Only one horizontal fraction (division) line is used, keeping the numerators and denominators aligned horizontally. For example, say we have the following multiplication to perform:

$$2 \cdot \frac{7}{9} \cdot \frac{5}{11}$$

Fraction bar notation would look like this:

2	7	5
	9	11

Fraction bars can usually eliminate confusing fractions-within-fractions. Fractions in the numerator are simply moved down onto the main fraction line. The following can be rewritten accordingly:

$$\frac{\left(\dfrac{3}{4}\right)}{8} \cdot \frac{2}{16}$$

In fraction bars the expression appears as

3		2
4	8	16

Fractions in the denominator are inverted and placed on the main fraction line (since dividing by a fraction is the same as inverting and multiplying). Thus the following can be set up in fraction bar notation:

$$\frac{7}{\left(\dfrac{3}{4}\right)} \cdot \frac{23}{14}$$

The result is

7	4	23
	3	14

You can rearrange the factors above the line in any order or rearrange those *below* the line in any order. But you cannot move them across the horizontal fraction line. All the compartments (factors) need not be filled.

Fraction bar notation is especially useful for equations that involve multiple units. Consider the following example. The SI unit of time is the second (s) (Chapter 9). In order to find out how many seconds there are in a day, you would multiply 60 seconds per minute by 60 minutes per hour by 24 hours per day.

60 s	60 min	24 h
min	h	d

After eliminating those units that cancel, we multiply the remaining factors.

$$= \frac{60 \text{ s} \cdot 60 \cdot 24}{\text{d}}$$

$$= 86\ 400 \text{ s/d}$$

Thus there are 86 400 seconds per day.

2.7 SCIENTIFIC CALCULATORS

A scientific calculator greatly simplifies the use of SI, especially the handling of prefixes (Chapter 4). For some of the example problems, we will give the sequence of keystrokes for a common algebraic-entry scientific calculator. On these calculators, numbers and operator keys (+, −, ×, ÷, y^x) are generally entered in the same order as they appear on paper. For example, say we are to evaluate the expression

$$3 \cdot \frac{4}{5} \cdot 2^3$$

We key in the following:

3 ☒× 4 ☒÷ 5 ☒× 2 ☒y^x 3 ☒=

The answer 19.2 is displayed. A few calculators use reverse Polish notation (RPN), in which the operator key is pressed *after* the number it acts on. With an RPN calculator you would evaluate the same expression above by entering

3 ☒ENTER 4 ☒× 5 ☒÷ 2 ☒ENTER 3 ☒y^x ☒×

If you have an RPN calculator, refer to your manual for further details.

When using a calculator to solve a problem set up in fraction bars, simply multiply all the factors (compartments) above the line and divide all the factors (compartments) below the line. Assume we must evaluate

7.5	3.8	12
9.1	2.4	13

On an algebraic calculator, we enter

7.5 ☒× 3.8 ☒× 12 ☒÷ 9.1 ☒÷ 2.4 ☒÷ 13 ☒=

The (rounded) answer 1.2 is displayed.

Numbers in powers-of-ten format are entered with the "exponent" key (labeled ☒E, ☒Exp, ☒EE, or ☒EEX), which is calculator shorthand for "times ten to the…." In this book, the symbol ☒E is used for this key. Negative exponents are entered with the negative or "change-sign" key (labeled ☒+/−, ☒±, ☒(−), or ☒CHS), not the subtraction key. For example, to enter the number $2.5 \cdot 10^{-6}$, key in

2.5 ☒E ☒+/− 6

This will display on most calculators as 2.5E−6.

To raise a quantity to a power, use the "power" key (labeled ☒y^x or ☒a^b). Use the ☒π key to enter pi (≈3.14159). For example:

4	π	$(6.3 \cdot 10^{-3})^3$
3		

Enter

4 ☒× ☒π ☒× 6.3 ☒E ☒+/− 3 ☒y^x 3 ☒÷ 3 ☒=

The answer 1.0E−6 is displayed. We did not need to enter the written parentheses because the calculator recognizes a number entered in powers-of-ten format as a single entity.

Problem Set 2
SI Structure and Mathematical Conventions

1. Define quantity.

2. Define unit.

3. Quantity symbols are written in what letter style (italic, bold, or regular)?

4. What letter style is used for unit symbols?

5. List the seven SI base units, their symbols, and the quantities they measure.

6. The two SI angle units are classified as what kind of unit?

7. Units that are defined as products or quotients of the base units are classified as what kind of units?

8. What does "per" mean?

9. Write "kilogram per second" in symbols, three different ways.

10. Write "kilogram second" in symbols.

11. What is the term for derived units without special names, such as meter per second?

Indicate whether the following are quantities or units.

12. 300 meters

13. m

14. 273 K

15. m

16. kelvin

17. volume

18. What are the three groups of SI symbols that are capitalized? (See Appendix A, rule 2.)

Which is the correctly written version of each quantity below?

19. (A) 35mm   (B) 35 MM
    (C) 35 mm.   (D) 35 mm

20. (A) 18 kms   (B) 18 KM.
    (C) 18 km   (D) 18 km.

21. (A) 50 Newtons   (B) 50 N.
    (C) 50 n   (D) 50 newtons

22. (A) .5 meters   (B) $\frac{1}{2}$ m
    (C) 0.5m   (D) 0.5 m

23. (A) 22 kg (B) 22 kgs.
    (C) 22 Kg   (D) 22 kg

24. (A) 30 m/sec   (B) 30 mps
    (C) 30 m/s   (D) 30 meters/second

25. (A) 1.8 J·s   (B) 1.8 J-s
    (C) 1.8 J s   (D) 1.8 Joule seconds

26. (A) 15 MHertz   (B) 15 MHz
    (C) 15 megaHz   (D) 15 Mhz

27. (A) 3 m 50 cm   (B) 3.5 m
    (C) $3\frac{1}{2}$ m   (D) 3.5 m

Which way of writing the following numbers is preferred in SI?

28. (A) 38 569.23   (B) 38,569.23   (C) 38.569,23

29. (A) 0.00279   (B) 0,00279   (C) 0.002 79

Rewrite in fraction-bar notation.

30. $$\frac{\left(\dfrac{5}{4}\right)}{9} \cdot \frac{1}{16} \cdot \frac{7}{\left(\dfrac{2}{3}\right)} \cdot 13$$

31. $$\frac{5.98 \cdot 10^{24} \text{ kg}}{\dfrac{4}{3}\pi\,(6.37 \cdot 10^6 \text{ m})^3}$$

Evaluate the following expression with a calculator.

32. $$\frac{\pi \quad\big|\quad 2.4 \cdot 10^{-4} \quad\big|\quad 9^5 \quad\big|}{\quad\big|\quad 1.8 \cdot 10^9 \quad\big|\quad 7.8 \quad\big|\quad 12}$$

CHAPTER 3

Precision and Rounding

3.1 SIGNIFICANT DIGITS

Measurements are never exact. The numbers in physical quantities are always approximate. The digits that were actually measured are called *significant digits*. In a correctly written quantity, all digits are significant except placeholding zeroes, which can occur at either end of the number.

• Leading zeroes are never significant. For example, the three zeroes that lead off in the quantity "0.002 meter" are just placeholders to show where the decimal point goes. They were not measured.

• Ending zeroes to the *left* of an (implied) decimal point are usually not significant, but you cannot be sure without other information. For example, most of the zeroes in the quantity "3000 years" are probably not significant. The actual, measured time might be anything from 2500 years to 3499 years. (A decimal point is understood at the end of the number even though it normally is not written).

Note that all other zeroes *are* significant, including ending zeroes to the right of the decimal point. For example, "30.0 m" means a measurement of precisely 30.0 meters—not 29.9 or 30.1 meters.

The *precision* of a quantity is the number of significant digits it contains. An *exact* number has infinite precision—an infinite number of significant digits. Only pure numbers, definitions, and counting numbers may be exact. A pure number is a number alone, without a unit, that is, not part of a quantity. Counting numbers are explained in Section 3.5. Measured quantities are *never* exact. The best measurements today have 10 or 12 significant digits, but such extreme precision is rare. Study the following examples:

Quantity	Number of significant digits
358.6 kg	4
0.03 m	1
0.0060 kg	2
5000 K	? (1, 2, 3, or 4)
200.0 m	4
40 s	? (1 or 2)

3.2 ROUNDING RULES

When working with quantities, you must round answers to the same precision as the original data or measurements. In other words, you should end up with the same number of significant digits or decimal places that you started with. Failure to round off is a serious error and can even be dishonest. There are four general rules for rounding:*

1. If the next digit (to be discarded) is 5 or greater, round up. If it is 4 or less, round down.

For example, 5.3478 rounded to two significant digits is 5.3, but 5.3521 rounded to two significant digits is 5.4.

2. When multiplying or dividing, round answers to the fewest significant digits found in any of the original data.

For example,

$$23.206 \cdot 1.8 \cdot 0.0300 = 1.3$$

The fewest significant digits was two (in the factor 1.8), so we rounded the answer to two significant digits.

3. When adding or subtracting, imagine the numbers arranged with the decimal points aligned vertically, as if adding by hand. The answer should have no significant digits farther right than there are in the least precise number.

For example (assuming the ending zeroes are not significant),

$$
\begin{array}{r}
134\,000 \\
23\,665.879 \\
+\ 267\,893. \\
\hline
425\,558.879 \quad \text{which rounds to } 426\,000
\end{array}
$$

The least precise number was 134 000, in which the significant digit farthest right is 4 (the thousands digit). So we rounded the calculated answer to the nearest thousand, or 426 000.

If all the terms have digits to the right of the decimal point we can express rule 3 more simply: round to the *fewest decimal places* found in any of the original quantities. Decimal places are the digits to the right of the decimal point.

4. Do not round off until you are finished.

While doing calculations, carry extra digits so that rounding errors do not accumulate. (Some loss of precision is inevitable when raising quantities to a power).

---

*From *American National Standard for Metric Practice* (see Bibliography).

Pure numbers, such as defined conversion factors and the factors in mathematical formulas, are not measured quantities. They may be exact and are therefore ignored when following the rules above. For example, the factor 4 in the formula for the area of a sphere ($A = 4\pi r^2$) is exact and is ignored when rounding.

Most scientific calculators will automatically round answers to the precision you specify. For most operations, use the SCI or ENG* display options to specify the number of significant digits. When adding or subtracting only, use the FIX display option to specify the number of decimal places.

3.3 ACCURACY

Precision tells you how precise a number is—how many significant digits or decimal places it has. *Accuracy,* on the other hand, tells you how "close to true" a number is. For example, the statement "a standard door is 2 m high" is completely accurate; a door is not 1 m or 3 m high. But it is not very precise—only one significant digit. The statement "a standard door is 2.03 m high" is more precise but *less* accurate, because doors can vary several centimeters owing to adustments for carpeting and the like.

As another illustration, suppose you have an instrument that gives readouts to five significant digits; it would be a very precise device. But the instrument might be broken or might need calibration. Thus the numbers that it gives would be precise but inaccurate. In contrast, a cheap ruler marked in

*On many inexpensive calculators, the ENG display does not round properly, in which case you must round by hand. But the ENG display is still a great advantage for handling prefixes; see Chapter 4.

centimeters is 100% accurate when you measure to the nearest centimeter, but it is not very precise.

3.4 ESTIMATING PRECISION

You must use common sense when estimating precision and rounding off. It is important not to imply more precision than is warranted. You should not try to interpolate between the lines on a ruler, gage, or analog meter dial. Digital instruments sometimes display more significant digits than they can accurately measure; check the specifications.

In practice, significant zeroes to the right of the decimal point are often omitted. A measurement of "8 meters" may mean 8.0 m or 8.00 m. A "2 liter" bottle of soda probably contains 2.00 liters. However, the non-SI labels on some products are traditional names or nominal values, not literal dimensions. For example, a "2 by 4" inch board actually measures about $1\frac{1}{2}$ by $3\frac{1}{2}$ inches. The diameter of a "1 inch" pipe varies with the type of pipe and is never quite 1 inch. This must be taken into account when converting non-SI quantities to SI, or vice versa.

3.5 COUNTING NUMBERS

The number of discrete objects is not a measurement; objects may be counted in exact whole numbers. You cannot have 1.3 chairs or 0.5 dog. You must use common sense and round off when appropriate. It would be incorrect to say that the population of a city is 886 645 people. Population constantly changes and we cannot count such a large number of persons exactly. We should round this number to 887 000 or 900 000 when reporting it.

Problem Set 3
Precision and Rounding

1. What is the term for a measured digit?

2. What term refers to the number of significant digits in a measurement?

3. What word describes a number with infinite precision?

4. Which zeroes in a number are *never* significant?

5. Which zeroes are *usually* not significant (although some may be)?

6. What term describes how "close to true" a quantity is?

Give the number of significant digits.

7. 89.2

8. 4001

9. 0.034

10. 650

11. 500.0

12. 72 000

13. 0.0040

14. 0.700 L

15. 702 km

16. 905.7 kg

17. 0.0009 m

18. 5800 K

19. 200.00 m/s

20. 0.3700 s

Round the calculated answers below to correct precision. Assume they are all measurements (approximations) and that ending zeroes to the left of the decimal point are not significant.

21. $3.44 \cdot 2 = 6.88$

22. $18.7 + 4.98 = 23.68$

23. $\dfrac{1504}{2.89} = 520.42$

24. $458\,000 + 659 - 480 + 20.73 = 458\,199.73$

25. $564\,980 \cdot 5.1 = 2\,881\,398$

26. $20.890 \div 5.000 = 4.1780$

27. $800 \cdot 1.2578 \div 153.21 \cdot (2.92 + 23) = 170.235$

CHAPTER **4**

Prefixes

4.1 ADVANTAGES OF PREFIXES

SI *prefixes* are short names and letter symbols for numbers. They are displayed later in this chapter and are summarized in Table 1 on the front inside cover. A prefix is attached in front of a unit name as a multiplying factor. For example, mega (M) means million (10^6), and thus megawatt (MW) is a simple way of expressing 1 000 000 watts.

Prefixes eliminate awkward placeholding (nonsignificant) zeroes and hard-to-pronounce powers of ten. They replace tongue-twisting names like billion, billionth, trillion, and trillionth, which sound alike, are often confused, and have a different meaning in most countries. In British English and nearly all other European languages, "billion" means 10^{12}, whereas in the United States it means 10^9. In most languages, "trillion" means 10^{18}, whereas in the United States it means 10^{12}. This confusion continues for all numbers larger than a million or smaller than a millionth. The SI prefixes, by contrast, are universal and easier to pronounce.

4.2 POWERS OF TEN

Because we have ten fingers, we use a base-ten or *decimal* number system for ordinary purposes. In such a system it is convenient to express huge or tiny numbers with powers of ten. When manipulating powers by hand, you will need to recall the exponent rules from algebra (see Table 4.1). A scientific calculator takes care of these automatically.

TABLE 4.1 Exponent Rules

$$10^{-a} = \frac{1}{10^a}$$

$$10^a \cdot 10^b = 10^{a+b}$$

$$\frac{10^a}{10^b} = 10^{a-b}$$

$$10^0 = 1$$

$$\left(10^a\right)^b = 10^{ab}$$

$$\sqrt[a]{10} = 10^{1/a}$$

$$(n \cdot m)^a = n^a \cdot m^a$$

4.3 THREE NOTATIONS

There are three common notations for expressing numbers. Most scientific calculators will display in any notation you select, automatically rounding to the precision you specify. Numbers may be entered in any notation, regardless of the display chosen.

• *Ordinary notation* (called FIX, Floating Point, or ALL on calculators) uses placeholding zeroes rather than powers of ten. If the numbers are quite large or small, ordinary notation is difficult to read, write, and manipulate. For example, the mean distance to the sun is 150 000 000 000 meters in ordinary notation. We cannot even pronounce this number without counting all the zeroes first. The precision is also unclear, since you do not know how many of the zeroes are significant (only the first is).

• *Scientific notation* (SCI on calculators) expresses numbers in the form $n.nn\ldots \cdot 10^a$. For example, the distance to the sun in scientific notation is $1.50 \cdot 10^{11}$ meters. The number $n.nn\ldots$ always has just one digit (1 through 9) to the left of the decimal point. There can be any number of digits to the right, depending on the precision. Since there are never any zeroes to the left of the decimal point, all digits are always significant in scientific notation. That is its main advantage.

• *Prefix notation* (called ENG for "engineering" on calculators) is like scientific notation except that the exponent is always a multiple of 3 (±3, 6, 9, 12, etc.). Those powers of ten have convenient names and letter symbols—the SI prefixes. For example, the distance to the sun in ENG notation is $150 \cdot 10^9$ meters, which we write more simply as 150 Gm (pronounced and spelled out "150 gigameters").

You may use any notation with SI but prefix notation is the most common because it is easiest to write, say, and remember, and it usually indicates the precision within one or two digits. Avoid mixing prefixes and powers of ten in the same quantity. That is confusing and redundant since the prefix *is* a power of ten, and comparison can be difficult. For example, write 10^5 m, not 10^7 cm. The kilogram is an exception (see Section 8.2).

4.4 PREFIX NAMES AND SYMBOLS

As shown in Table 4.2, the larger prefixes, mega and up, end in the letter *a* and have capital symbols. Those from kilo

TABLE 4.2 Prefixes and Their Equivalents

Symbol	Prefix	Factor	Ordinary notation	U.S. name	European name (if different)
Y	yotta	10^{24}	1 000 000 000 000 000 000 000 000	septillion	*quadrillion*
Z	zetta	10^{21}	1 000 000 000 000 000 000 000	sextillion	*trilliard*
E	exa	10^{18}	1 000 000 000 000 000 000	quintillion	*trillion*
P	peta	10^{15}	1 000 000 000 000 000	quadrillion	*billiard*
T	tera	10^{12}	1 000 000 000 000	trillion	*billion*
G	giga	10^{9}	1 000 000 000	billion	*milliard*
M	mega	10^{6}	1 000 000	million	
k	kilo	10^{3}	1 000	thousand	
h	hecto	10^{2}	100	hundred	
da	deka	10^{1}	10	ten	
		10^{0}	1	one	
d	deci	10^{-1}	0.1	tenth	
c	centi	10^{-2}	0.01	hundredth	
m	milli	10^{-3}	0.001	thousandth	
µ	micro	10^{-6}	0.000 001	millionth	
n	nano	10^{-9}	0.000 000 001	billionth	*thousand millionth*
p	pico	10^{-12}	0.000 000 000 001	trillionth	*billionth*
f	femto	10^{-15}	0.000 000 000 000 001	quadrillionth	*thousand billionth*
a	atto	10^{-18}	0.000 000 000 000 000 001	quintillionth	*trillionth*
z	zepto	10^{-21}	0.000 000 000 000 000 000 001	sextillionth	*thousand trillionth*
y	yocto	10^{-24}	0.000 000 000 000 000 000 000 001	septillionth	*quadrillionth*

down have lowercase symbols and mostly end in the letter *o*. The symbol for micro is µ, the lowercase Greek letter *m* (pronounced *MOO* or *MEW*), because the lowercase roman "m" is already overworked as the symbol for meter and milli. SI symbols were carefully chosen to avoid ambiguity. However, there are two symbols with dual meanings: lowercase "m," just mentioned, and capital "T," which means both tera (10^{12}) and tesla (a specialized unit that measures magnetic flux density).

Prefix names were adapted from various languages, although the newer ones have been modified almost beyond recognition. The following etymologies may help you remember them. All prefixes are stressed on the first syllable.

Prefix	Pronunciation	Origin
yotta	[*YOTE-uh* or *YOTT-uh*]	"eight" [$(10^3)^8$] (Latin)
zetta	[*ZETT-uh*]	"seven" [$(10^3)^7$] (Latin)
exa	[*EX-uh*]	"six" [$(10^3)^6$] (Greek)
peta	[*PET-uh*]	"five" [$(10^3)^5$] (Greek)
tera	[*TAIR-uh*]	"monster" (Greek)
giga	[*JIG-uh* or *GIG-uh*]	"giant" (Greek)

mega	[*MEG-uh*]	"big" (Greek)
kilo	[*KILL-oh*]	"thousand" (Greek)
hecto	[*HECK-toe*]	"hundred" (Greek)
deka	[*DECK-uh*]	"ten" (Greek)
deci	[*DESS-ih*]	"tenth" (Latin)
centi	[*SENT-ih*]	"hundredth" (Latin)
milli	[*MILL-ih*]	"thousandth" (Latin)
micro	[*MIKE-roe*]	"small" (Greek)
nano	[*NAN-oh*]	"dwarf" (Greek)
pico	[*PEEK-oh*]	"tiny bit" (Spanish)
femto	[*FEM-toe*]	"fifteen" (Dano-Norwegian)
atto	[*AT-toe*]	"eighteen" (Dano-Norwegian)
zepto	[*ZEP-toe*]	"seven" [$(10^3)^{-7}$] (Latin)
yocto	[*YOCK-toe*]	"eight" [$(10^3)^{-8}$] (Latin)

Nano (10^{-9}) may remind you of "NINE-oh." Note that *t*era (T); its exponent, *t*welve; and its U.S. name, *t*rillion, all begin with the letter *t*.

4.5 CHOOSING PREFIXES

Most of the 20 prefixes—and those most commonly used—are multiples of a thousand (10^3). These regular prefixes are shown without shading in Table 4.2 (and in Table 1). They multiply or divide units in steps of 1000 as you go up or down the prefix ladder. For example, a kilometer is 1000 meters, a megameter is 1000 kilometers, a gigameter is 1000 megameters, and so on. In symbols,

$$
\begin{aligned}
1000 \text{ m} &= \text{km} \\
1000 \text{ km} &= \text{Mm} \\
1000 \text{ Mm} &= \text{Gm}
\end{aligned}
$$

Going down the scale, a millimeter is a thousandth of a meter, a micrometer is a thousandth of a millimeter, a nanometer is a thousandth of a micrometer, and so on. In symbols,

$$
\begin{aligned}
0.001 \text{ m} &= \text{mm} \\
0.001 \text{ mm} &= \mu\text{m} \\
0.001 \text{ } \mu\text{m} &= \text{nm}
\end{aligned}
$$

A unit with a prefix attached is called a *multiple* of the unit.* It does not form a separate unit. You would not consider 1000 feet a different unit from feet. Likewise kilometer, which means 1000 meters, is not a different unit from meter.

So that quantities will be easy to read and will indicate the approximate precision, try to use a prefix that yields a number between 0.1 and 1000. For example, 8 km is preferable to 8000 m; 20 g is preferable to 0.020 kg. However, quantities tabulated in a column should all have the same prefix, for ease in comparison, even if the range is large. For areas and volumes it may also be necessary to use several placeholding zeroes (see Chapters 6 and 7).

Never use more than one prefix with a unit symbol. For example, write nanometer (nm), not "millimicrometer" (mμm). In a compound unit, use a prefix only with the left-hand or numerator unit. Otherwise, comparing quantities will be difficult. For example, use kilonewton meter (kN·m), not newton millimeter (N·mm). Use watt per square meter (W/m^2), not watt per square centimeter (W/cm^2). However, the kilogram and density units are exceptions to this rule (see Chapter 8).

4.6 CHANGING PREFIXES BY HAND

You will often want to switch prefixes to simplify an expression and eliminate unnecessary, nonsignificant (placeholding) zeroes. Changing prefixes should not be called "converting units" since no arithmetic is involved—you just move the decimal point. In most cases, a scientific calculator will do this automatically if it is set to ENG display (see Section 4.8). To change prefixes that are multiples of 1000 by hand, go up or down Table 1 or Table 4.2 and "jump" the decimal point *three places* for each unshaded prefix you pass.

*Prefixes for multiples less than 1 are technically called *submultiples,* but the term is rarely used.

Ignore (skip) the shaded prefixes, which are not multiples of 1000. Move the decimal point *left* to switch to a larger prefix (up the table). Move it *right* to switch to a smaller prefix (down the table). Remember: "*Left* for *Larger*, *Right* for *Smaller*." Be sure to include 10^0 (no prefix) as a step if you pass it. Units attached to the prefix are just carried along, as shown in the examples below. The arrows demonstrate how to jump the decimal point.

- 7 200 m = 7.2 km   (kilometers)

We moved the decimal point 1 jump (3 places) left and thus stepped up 1 prefix larger, from no prefix to kilo.

- 0.015 s = 15 ms   (milliseconds)

We moved the decimal point 1 jump right and thus stepped *down* 1 prefix smaller, from no prefix to milli.

- 5 300 000 mm = 5.3 km   (kilometers)

We moved the decimal point 2 jumps (6 places) left and thus stepped up 2 prefixes larger, from milli to kilo.

- 300 000 km/s = 300 Mm/s   (megameters per second)

We moved the decimal point 1 jump left and thus stepped up 1 prefix larger, from kilo to mega.

Numbers spelled out in words are treated like the powers of ten they represent:

- 75 million kilometers = 75 Gm   (gigameters)

Million is 10^6, or 2 jumps larger (1000 · 1000), so we stepped up 2 prefixes larger, from kilo to giga.

- 3.1 millionths of a millisecond = 3.1 ns
  (nanoseconds)

Millionths is 10^{-6}, or 2 jumps smaller, so we stepped down 2 prefixes smaller, from milli to nano.

For areas and volumes (multiples of the square and cubic meter), changing prefixes is a bit more complicated than just described, because the prefix must be squared or cubed. See Chapters 6 and 7.

4.7 CHANGING SCIENTIFIC NOTATION TO A PREFIX, BY HAND

If a quantity is in scientific notation, you may be able to substitute an equivalent prefix directly.

- $1.5 \cdot 10^3$ W = 1.5 kW   (kilowatts)

However, if the exponent is not a multiple of three (± 3, ± 6, ± 9, etc.), you will have to juggle the decimal point and exponent to put the quantity into engineering notation. By hand, the procedure is to multiply or divide the power of ten by 10 or 100 to get an appropriate exponent. Then you must do the opposite operation (divide or multiply by 10 or 100) to the number on the left side of the multiplication sign, so that the whole expression remains unchanged. In practice, you do this by moving the decimal point the same number (of places) that you increase or decrease the exponent. If you *increase* the exponent, move the decimal point *left*. If you *decrease* the exponent, move the decimal point *right*.

- $4.3 \cdot 10^4$ m $= 43 \cdot 10^3$ m $= 43$ km

We decreased the exponent by 1 (divided the power of ten by 10), so we had to *multiply* the left side by 10 (by moving the decimal point 1 place right).

- $6.8 \cdot 10^{-4}$ s $= 680 \cdot 10^{-6}$ s $= 680$ μs

We divided the right side by 100 (decreasing the exponent by 2), so we had to multiply the left side by 100 (by moving the decimal point 2 places right).

- $2.9 \cdot 10^5$ J $= 290 \cdot 10^3$ J $= 290$ kJ

Increasing the exponent by 1 (to 6) would have given us an awkward leading decimal point (0.29), so we *decreased* the exponent by 2 (to 3) and moved the decimal point 2 places right.

These manipulations can be a bother and a source of mistakes. The simple solution is to use a scientific calculator set to ENG display.

4.8 HANDLING PREFIXES ON CALCULATORS

In most cases, a scientific calculator that is set to the ENG display will change prefixes for you. For all but the simplest problems, you may wish to leave the calculator set on ENG. The best models will even display SI letter symbols and will let you enter them on the keypad. On any scientific calculator, you can enter prefixes as powers of ten, using the exponent key $\boxed{E}$, as described in Section 2.7. Remember that a prefix is really part of the number, although the prefix is physically attached to the unit. For example, to enter 3.6 mm, which is short for $3.6 \cdot 10^{-3}$ m, key in

$$3.6 \quad \boxed{E} \quad \boxed{+/-} \quad 3$$

When entering areas (multiples of square meter) you must square the prefix. When entering volumes (multiples of cubic meter) you must cube the prefix. See Chapters 6 and 7 for details.

There are three basic options for dealing with prefixes in calculations:

- If all your data have the same prefix, it may be easier to ignore it. Your answer will have the same prefix. For example, if you are adding lengths in millimeters and you want an answer in millimeters, it would be a waste of keystrokes to enter the "milli" ($\boxed{E}$ $\boxed{+/-}$ 3) in your calculator. However, you may need to simplify the final answer to get rid of excess placeholding zeroes, as described in Section 4.6.

If your data have different prefixes, you cannot ignore them. You cannot add millimeters to centimeters, or multiply meters by kilometers. You have two choices:

- You can move the decimal point(s) to make all the prefixes the same or eliminate them altogether. Often this can be done mentally as you enter the numbers in the calculator. For example, you might enter 150 mm as 0.15 m, eliminating the prefix "milli." Your answer will have the same prefix as the data you entered.

- You can enter all the prefixes in your calculator with the exponent key, $\boxed{E}$. This method may require more keystrokes, but it is the best method for complex problems because you are less likely to make a mistake. You just enter every prefix methodically. Your raw answer will not have a prefix, but you may wish to simplify it by substituting one.

The example problems in Part II of this book illustrate these three approaches.

4.9 PREFIXES THAT ARE NOT MULTIPLES OF 1000

The prefixes hecto, deka, deci, and centi (shaded on Tables 1 and 4.2) should be avoided where practical because they needlessly complicate SI by making calculation, comparison of data, and prefix switching more difficult. However, they are commonly used when measuring areas and volumes (Chapters 6 and 7); and "centi" is a popular prefix for measuring length (see Section 5.2).

Problem Set 4
Prefixes

1. Give three advantages of prefixes.

2. Which of the three notations is usually easiest to write and pronounce?

3. In which notation are all digits always significant?

Prefix puns (substitute a prefix for the number).
Example: 1000 flies = kilofly.

4. 10^{12} bulls

5. 1 000 000 000 000 000 dogs

6. million bucks

7. 0.001 pede

8. 10^{18} rays

9. 10^{-18} boy

10. 0.000 000 000 001 boo

11. Two SI symbols have dual meanings. List both, along with their meanings.

12. What is the U.S. name for 10^{12}?

13. What is the British and common European name for 10^{12}?

14. What do you call a unit with a prefix attached?

Simplify the following quantities by changing prefixes. Eliminate excess nonsignificant zeroes, leading zeroes, powers of 10, and U.S. number names. Assume ending zeroes to the left of the decimal point are not significant. (Do not use a calculator.)

15. 0.001 m

16. 1000 m

17. $\frac{1}{1000}$ mm

18. 1 000 000 ms

19. 1000 kJ

20. 0.001 mm

21. 0.001 μm

22. 1000 kg

23. 0.001 Gm

24. 1 000 000 mm

25. 0.000 001 km

26. 8700 m

27. 53 000 W

28. 0.017 W

29. 0.000 060 s

30. 0.03 g

31. 430 000 kJ

32. 700 000 MN

33. $6.3 \cdot 10^{-9}$ s

34. 2350 mm

35. $14.3 \cdot 10^{-6}$ Pa

36. 1800 ms

37. 2500 kJ

38. 0.0023 nJ

39. 0.07 mm

40. 150 000 000 km

41. 0.035 kK

42. 1500 mg

43. 7900 MHz

44. 0.09 A

45. 0.005 mV

46. 35 thousand kilometers

47. 591 million kilometers

48. 14 millionths of a watt

49. 8.6 billion (U.S.) kilowatts

50. 90 millionths of a millimeter

51. 58 billionths (U.S.) of a millisecond

52. $4.5 \cdot 10^5$ m

53. $6.2 \cdot 10^{-2}$ N

54. $0.038 \cdot 10^5$ s

55. $0.026 \cdot 10^7$ W

Which of the following expressions is preferred in SI?

56. (A) 2000 mg   (B) 2 g   (C) 0.002 kg

57. (A) kJ/m^2 (B) J/mm^2 (C) kJ/cm^2

PART II

Quantities and Their Units

CHAPTER **5**

Length

5.1 THE METER

The quantity of length (*l*) has many names, such as distance, dimension, displacement, depth, width, thickness, diameter, radius, circumference, and size. But they are all synonyms for, or kinds of, length. In the United States, length is measured with a great variety of non-SI units. But in SI, all lengths are measured with the same unit—the meter (m). *Meter* comes from the Greek word *metron*, meaning "measure." (In all English-speaking countries except the United States, it is spelled *metre*.) The meter was originally based on the size of Earth (see Figure 5.1). The distance from the equator to the North Pole was arbitrarily defined as 10 million meters (10 megameters, or 10 Mm).

Earth is not quite spherical, and slight errors were made in the original survey. But for practical purposes you can think of Earth as being 40 megameters (40 Mm) in circumference. In other words, a meter is about 1/40 000 000 of Earth's circumference.

Until 1960 the meter was officially defined by the length of a platinum bar kept in a vault in France. However, that was not precise enough for modern needs and there was concern that the bar might slowly change dimension or be damaged in a calamity. Today, like all the base units except the kilogram, the meter is defined by unvarying natural phenomena that can be measured anywhere in the universe. The modern definition of the meter is based on the speed of light, a constant that can be determined with great precision.

- A *meter* is the distance light travels in a vacuum (empty space) in exactly 1/299 792 458 of a second.

In other words, light travels at exactly 299 792 458 meters per second, by definition. For most purposes we can round this value to 300 Mm/s. The speed of light is the fastest possible speed and a very important quantity in nature. Note that we have not really changed the size of a meter since it was originally defined in 1790; we have just defined it more precisely.

5.2 MULTIPLES OF METER

A quick way to get comfortable with the meter is to learn an example for each multiple, like those in Figure 5.2 and in Table 5.1. A mental image of such multiples lets you decide whether a calculated answer is reasonable. Be sure to pronounce kilometer "*KILL-oh-meter*" (not "*kil-LOM-muh-ter*"). SI multiples are always stressed on the first syllable so that the sound of the prefix is not lost.

For everyday purposes, the prefix centi (c = 10^{-2} = 0.01) is often used with the meter, although it is not a multiple of 1000. Because a centimeter (cm) equals 0.01 m, then 100 cm equals 1 m and 10 mm equals 1 cm. To change the prefix on centimeter you must therefore move the decimal point 1 or 2 places instead of the usual 3 (Section 4.6). The calculator will not automatically do it for you. For example,

$$352 \text{ cm} = 3.52 \text{ m}$$
$$0.65 \text{ m} = 65 \text{ cm}$$
$$2.4 \text{ cm} = 24 \text{ mm}$$
$$760 \text{ mm} = 76 \text{ cm}$$

For technical use, the millimeter is preferred over the centimeter.

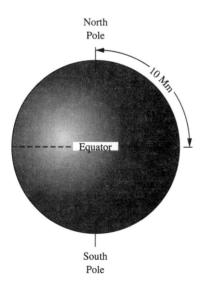

North
Pole

10 Mm

Equator

South
Pole

FIGURE 5.1 Original definition of the meter.

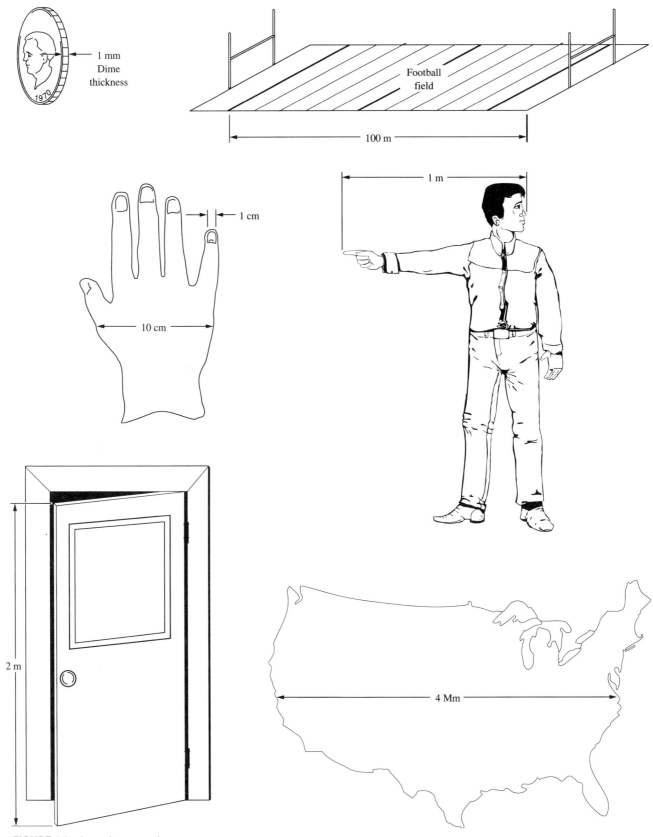

FIGURE 5.2 Length comparisons.

TABLE 5.1 Multiples of the Meter

Symbol	Name	Equivalent	Examples
Ym	yottameter	10^{24} m	the universe (estimated at 100–200 Ym)
Zm	zettameter	10^{21} m	galaxies (Milky Way diameter ≈ 1 Zm)
Em	exameter	10^{18} m	distance of farther stars (within our galaxy)
Pm	petameter	10^{15} m	closer stars (Alpha Centauri ≈ 40 Pm distant)
Tm	terameter	10^{12} m	the solar system (Pluto's orbit ≈ 12 Tm diameter)
Gm	gigameter	10^{9} m	the sun's distance from Earth (≈ 150 Gm)
Mm	megameter	10^{6} m	planets (Earth's circumference ≈ 40 Mm)
km	kilometer	10^{3} m	mountains (elevation of Mt. Everest ≈ 9 km)
m	meter	*base unit*	arm's length; waist high (≈ 1 m)
cm	centimeter	10^{-2} m	width of a fingernail (≈ 1 cm)
mm	millimeter	10^{-3} m	dime thickness (≈ 1 mm)
μm	micrometer	10^{-6} m	bacteria (1–10 μm)
nm	nanometer	10^{-9} m	viruses (20–400 nm)
pm	picometer	10^{-12} m	atoms (100–600 pm)
fm	femtometer	10^{-15} m	protons or neutrons (≈ 2 fm)

Example Problem 5.1. The sun is 150 Gm from Earth, and the moon is 384 Mm from Earth. How many times farther is the sun than the moon?

Solution. To compare these two lengths we divide the larger by the smaller.

$$\frac{150 \text{ Gm}}{384 \text{ Mm}} = \frac{150 \cdot 10^{9} \text{ m}}{384 \cdot 10^{6} \text{ m}} = 390$$

The units (meter) canceled, so the answer is a pure number. To solve this problem with a calculator, we enter

150 $\boxed{\text{E}}$ 9 $\boxed{\div}$ 384 $\boxed{\text{E}}$ 6 $\boxed{=}$

We rounded the answer to two significant digits because we realized that the distances vary somewhat (the orbits of the moon and Earth are not perfect circles).

Non-SI Length Units Used in the United States

angstrom	yard	mile	light year	screw number
mil	fathom	survey mile	parsec	nail penny
inch (decimal)	rod	nautical mile	American wire gage	shotgun gauge
inch (fractional)	link	point	W&M gage	men's shoe size
hand	chain	pica	USS gage	women's shoe size
foot	furlong	agate	B&S gage	children's shoe size
survey foot	cable	astronomical unit	drill number	

Problem Set 5
Length

1. What is the approximate circumference of Earth in SI units (to two significant digits)?

2. What is the exact modern definition of a meter?

3. What is the speed of light, rounded to three significant digits?

4. Express 56 cm in millimeters.

5. Express 9.3 cm in meters (without a prefix).

6. Express 16 mm in centimeters.

7. Which pronunciation of kilometer is correct?

    (A) *kil-LOM-muh-ter*   (B) *KILL-o-meter*

8. What is the height of a standard door?

    (A) 1200 mm   (B) 130 cm
    (C) 2 m   (D) 3 m

9. What is the approximate distance from New York City to San Francisco?

    (A) 400 km   (B) 4 Mm
    (C) 40 Mm   (D) 400 000 m

10. About how many times larger in diameter is a hydrogen atom (160 pm) than its nucleus, a proton (1.6 fm)?

11. About how many times farther is the nearest star, Alpha Centauri (40 Pm), than our sun (150 Gm)?

12. A map has a scale of 1/500 000 (that is, distances on the map are 1/500 000 of the corresponding distances on the ground). If two towns are 75 km apart, how far apart do they appear on the map?

Convert the following quantities to SI, using the definitions in Appendix B (see Sections B.1 and B.2 for conversion instructions). Express answers in correct precision and simplest form, using an appropriate prefix.

13. 45.6 miles (international)

14. 45.6 miles (nautical)

15. 14 feet $5\frac{1}{4}$ inches

16. 85 light-years

CHAPTER 6

Area

6.1 THE SQUARE METER

Area *(A)* is the quantity of surface (including imaginary surfaces). Since a surface has two dimensions we define area as length squared $(A = l^2)$. The SI unit of area is therefore the square meter (m^2). The word *square* should always be pronounced first. For example, 2 m^2 is pronounced "2 square meters." By contrast, "2 meters squared" means $(2\ m)^2$, which is $2\ m \cdot 2\ m = 4\ m^2$, or 4 square meters.

As far as prefixes are concerned, the square meter is the most difficult SI unit. It is the only SI unit that *never* steps by a thousand (10^3) per prefix. Prefixes on the square meter are tricky because the prefix is squared along with the unit. For example, a square kilometer, symbol km^2, means $(km)^2$, *not* $k(m^2)$. In other words, a square kilometer is equivalent to a square that is 1 kilometer, or 1000 m, on a side. Its area is therefore

$$1000\ m \cdot 1000\ m = 1\ 000\ 000\ m^2$$

We get the same answer more simply using powers of ten.

$$km^2 = (10^3\ m)^2 = 10^6\ m^2$$

Note that, although "kilo" means thousand, a square kilometer is *not* the same as a thousand square meters (1000 squares, each 1 meter on a side). When using a prefix with square meter, you cannot simply substitute the equivalent power of ten. The prefix must be squared. For prefixes that are multiples of 1000, the square meter therefore steps by a million (10^6) per prefix instead of the usual thousand (10^3).

6.2 PREFIXES HECTO AND CENTI

To avoid some of the inconvenient million-steps just described and the large number of placeholding zeroes that can result, the prefixes hecto (100) and centi (0.01) are commonly used with square meter.

$$hm^2 = (100\ m)^2 = 10^4\ m^2 = 10\ 000\ m^2$$
$$cm^2 = (10^{-2}\ m)^2 = 10^{-4}\ m^2 = 0.0001\ m^2$$

Unfortunately, we cannot make the square meter step by a thousand per prefix because the square root of a thousand is not a multiple of ten. When changing prefixes on square meter, you must therefore move the decimal point 2, 4, or 6 places per prefix instead of the usual 3.

A square hectometer (hm^2) is commonly called a *hectare* (ha), a name left over from an obsolete version of the metric system. A hectare is therefore equivalent to a square 100 m by 100 m (= 10 000 m^2) or is about the area of two football fields. There are 100 ha (or 100 hm^2) in a square kilometer (km^2).

6.3 MULTIPLES OF SQUARE METER

The prefixes normally used with square meter are summarized in Table 6.1. When changing prefixes on areas, you may find it easier to use this table instead of the prefix chart on the inside front cover. For areas larger than square megameters or smaller than square millimeters, prefixes are usually impractical because they require too many placeholding zeroes. If you should need to measure such huge or tiny areas, use scientific notation instead. It helps to picture multiples of the square meter as nesting squares, three of which are illustrated in Figure 6.1. The examples in Figure 6.2 will help you get a feel for multiples of the square meter.

Table 6.1 Common Multiples of Square Meter

Symbol	Name	Equivalent
Mm^2	square megameter	$10^{12}\ m^2 = 10^6\ km^2$
km^2	square kilometer	$10^6\ m^2 = 100\ ha = 100\ hm^2$
hm^2 (ha)	square hectometer (hectare)	$10^4\ m^2 = 10\ 000\ m^2$
m^2	square meter	$10^0\ m^2 = 10\ 000\ cm^2$
cm^2	square centimeter	$10^{-4}\ m^2 = 100\ mm^2$
mm^2	square millimeter	$10^{-6}\ m^2$

6.4 CHANGING PREFIXES

When changing prefixes on the square meter, move the decimal point the correct number of places, referring to Table 6.1. The following examples illustrate the procedure.

- $35\ 000\ m^2 = 3.5\ hm^2 = 3.5\ ha$

To get rid of the nonsignificant zeroes, we need to move the decimal point left and change to a larger prefix. The next common multiple larger than square meter is square hectometer

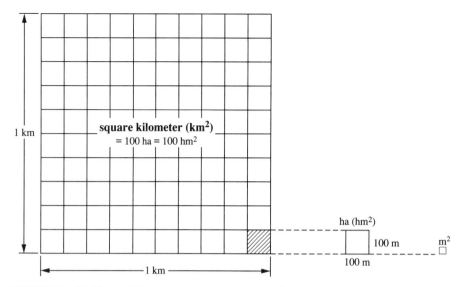

FIGURE 6.1 Multiples of the square meter make nesting squares.

(also called hectare), which is 10^4 square meters. So we moved the decimal point 4 places left (instead of the usual 3).

- 500 ha = 5 km^2

The next multiple larger than hectare is square kilometer, which is 100 ha, so we moved the decimal point 2 places left.

- 0.0035 m^2 = 35 cm^2

The next common multiple *smaller* than square meter is square centimeter, which equals 10^{-4} m^2, so we moved the decimal point 4 places *right*.

- 0.4 ha = 4000 m^2

The next common multiple smaller than hectare is square meter. A hectare is 10^4 m^2, so we moved the decimal point 4 places right.

- 12 million square kilometers = 12 Mm^2

The next multiple larger than square kilometer is square megameter, which equals 10^{12} m^2, or 10^6 (a million) km^2.

You may wish to assign a prefix to an area in scientific notation *without* referring to Table 6.1. First find the *square root* of the power of ten by dividing the exponent by 2. Then substitute a prefix for the root.

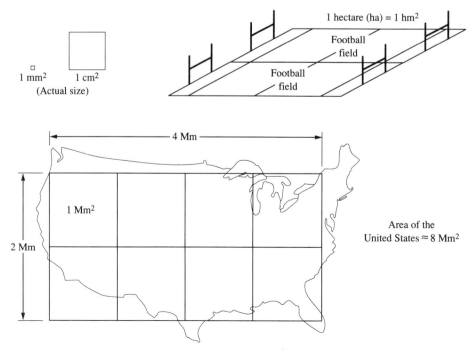

FIGURE 6.2 Area comparisons.

- $3.6 \cdot 10^4 \text{ m}^2 = 3.6 \cdot (10^2 \text{ m})^2 = 3.6 \text{ hm}^2 = 3.6 \text{ ha}$

We substituted the prefix hecto (h) for 10^2 and the alternative name hectare (ha) for square hectometer.

You may have to juggle the decimal point first to give a root with a prefix name (such as 10^2, 10^3, or 10^6). The process has been described in Section 4.7.

- $7.6 \cdot 10^{-3} \text{ m}^2 = 76 \cdot 10^{-4} \text{ m}^2 = 76 \cdot (10^{-2} \text{ m})^2 = 76 \text{ cm}^2$

We substituted the prefix centi (c) for 10^{-2}. In this case, setting the calculator to the ENG display did not help. Area is the *only* quantity for which the ENG display will not always give you a power of ten suitable for assigning a prefix.

Example Problem 6.1. What is the area of a rectangle 50 cm by 4 m?

Solution. We cannot multiply centimeters by meters; they have different prefixes. So we change 50 cm to 0.5 m and thereby eliminate the prefix.

$$0.5 \text{ m} \cdot 4 \text{ m} = 2 \text{ m}^2$$

Example Problem 6.2. What is the area of a rectangle 2.2 cm by 1.8 mm?

Solution. We cannot multiply centimeters by millimeters. We decide to eliminate the prefixes by entering them all in the calculator, rather than by moving decimal points. The answer will then be in square meters, without a prefix. We key in

$$2.2 \boxed{\text{E}} \boxed{+/-} 2 \boxed{\times} 1.8 \boxed{\text{E}} \boxed{+/-} 3 \boxed{=}$$

The rounded, displayed answer is 40E-6, which means $40 \cdot 10^{-6} \text{ m}^2$. We simplify this to 40 mm² by recalling Table 6.1. Alternatively, we could take the square root of 10^{-6}, dividing the exponent by 2 to get 10^{-3}, and then substitute the prefix milli (m) for the 10^{-3}.

Example Problem 6.3. What is the area of a circular irrigated field 805 m in diameter?

Solution. The area of a circle is πr^2. We must divide the diameter by 2 to get the radius. The given data is in meters so the answer will be in square meters. On the calculator we enter

$$\boxed{\pi} \boxed{\times} \boxed{(} 805 \boxed{\div} 2 \boxed{)} \boxed{y^x} 2 \boxed{=}$$

The rounded, displayed answer is 509E3, which means $509 \cdot 10^3 \text{ m}^2$. To express the number so it can be simplified with a prefix, we move the decimal point 1 place left and increase the exponent by 1, as in Section 4.7. The result is $50.9 \cdot 10^4 \text{ m}^2$, which simplifies to 50.9 hm², or 50.9 ha (Table 6.1).

Non-SI Area Units Used in the United States

acre (U.S. survey)	circular inch	square mile	square rod	township
acre (commercial)	square foot	square mil	square chain	barn
square inch	square yard	circular mil	section	

Areas of Common Figures

Parallelogram (including rectangle and square) $A = bh$ Triangle $A = \frac{1}{2} bh$ Circle $A = \pi r^2$ Sphere $A = 4\pi r^2$

Problem Set 6
Area

1. What is the quantity of surface called?

2. Give the alternative name and symbol for square hectometer (hm^2).

Express the following multiples as a power of ten of the square meter. *Example:* $km^2 = 10^6 \ m^2$.

3. square hectometer (hm^2)

4. square centimeter (cm^2)

5. square millimeter (mm^2)

6. hectare (ha)

Simplify the following quantities by changing prefixes. Include any alternative names.

7. 10 000 cm^2

8. 100 ha

9. 45 000 m^2

10. 0.0079 m^2

11. 900 mm^2

12. 7100 ha

13. 800 hm^2

14. 6 400 000 km^2

15. 0.32 km^2

16. 0.008 ha

17. 0.59 cm^2

18. 180 000 m^2

19. $71 \cdot 10^3 \ m^2$

20. $3.5 \cdot 10^{-3} \ m^2$

Calculate the area of the following. Express answers in the simplest form by using an appropriate prefix and correct precision. Assume ending zeroes to the left of the decimal point are not significant.

21. A football field (91 m by 49 m)

22. A rectangle 56 cm by 3.5 m

23. A rectangle 55 mm by 75 mm

24. A square field 200 m on a side

25. A circle of radius 177 cm

26. A circular field 290 m in diameter

27. The surface of a basketball (radius 12 cm)

28. Earth's surface (radius 6.4 Mm)

CHAPTER 7

Volume

7.1 THE CUBIC METER

Volume (V) is the quantity of space. Since a space has three dimensions we define volume as length cubed ($V = l^3$). The SI unit of volume is therefore the cubic meter (m^3). The word *cubic* should always be pronounced first. For example, 2 m^3 is pronounced "2 cubic meters." By contrast, "2 meters cubed" means $(2 \text{ m})^3$, which is $2 \text{ m} \cdot 2 \text{ m} \cdot 2 \text{ m} = 8 \text{ m}^3$, or 8 cubic meters.

Prefixes on the cubic meter are a bit tricky because the prefix must be cubed along with the unit (but they are not as difficult as those on the square meter). For example, a cubic kilometer, symbol km^3, means $(km)^3$, *not* $k(m^3)$. In other words, a cubic kilometer is equivalent to a cube that is 1 kilometer, or 1000 m, on each side. Multiplying the dimensions of this cube gives us a volume of

$$1000 \text{ m} \cdot 1000 \text{ m} \cdot 1000 \text{ m} = 1\,000\,000\,000 \text{ m}^3$$

We get the same answer more simply using powers of ten.

$$km^3 = (10^3 \text{ m})^3 = 10^9 \text{ m}^3$$

Note that, although "kilo" means thousand, a cubic kilometer is *not* the same as a thousand cubic meters (1000 cubes, each 1 meter on a side). When using a prefix with cubic meter, you cannot simply substitute the equivalent power of ten. It is very important that the *prefix be cubed*.

7.2 PREFIXES HECTO, DEKA, DECI, AND CENTI

Since huge billion-steps between prefixes would be inconvenient, the prefixes that are not multiples of 1000 are used with cubic meter. When cubed, they restore the usual steps of a thousand (10^3) for the most common multiples (between cubic millimeter and cubic kilometer).

cubic kilometer
$$= (10^3 \text{ m})^3 = 10^9 \text{ m}^3 = 1\,000\,000\,000 \text{ m}^3$$
cubic hectometer
$$= (10^2 \text{ m})^3 = 10^6 \text{ m}^3 = 1\,000\,000 \text{ m}^3$$
cubic dekameter
$$= (10^1 \text{ m})^3 = 10^3 \text{ m}^3 = 1000 \text{ m}^3$$
cubic decimeter
$$= (10^{-1} \text{ m})^3 = 10^{-3} \text{ m}^3 = 0.001 \text{ m}^3$$
cubic centimeter
$$= (10^{-2} \text{ m})^3 = 10^{-6} \text{ m}^3 = 0.000\,001 \text{ m}^3$$
cubic millimeter
$$= (10^{-3} \text{ m})^3 = 10^{-9} \text{ m}^3 = 0.000\,000\,001 \text{ m}^3$$

Note that deka is the only prefix with a two-letter symbol: da. In many languages it is spelled "deca." Do not confuse deka (da = 10) with deci (d = 0.1). Deka is rarely used.

7.3 ALTERNATIVE NAMES

A further complication of the cubic meter is that some of the prefixed multiples have alternative names and symbols. These terms are holdovers from an old version of the metric system. Although not coherent, they are approved for use with SI because they are so widespread and because they are easier to pronounce. You might think of them as "nicknames."

Alternative name	Formal SI name
liter (L)*	cubic decimeter (dm^3)
milliliter (mL)	cubic centimeter (cm^3)
microliter (µL)	cubic millimeter (mm^3)

No prefix other than milli or micro should be used with liter. In all English-speaking countries except the United States, liter is spelled "litre."

7.4 MULTIPLES OF CUBIC METER

It helps to picture multiples of the cubic meter as nesting cubes, three of which are illustrated in Figure 7.1. Note that as you go from one cube to the next larger, the sides increase *ten* times but the volume increases a *thousand* times (because $10^3 = 1000$). A thousand cubic centimeters (milliliters) makes a cubic decimeter (liter), a thousand cubic decimeters (liters) makes a cubic meter, and so on.

The prefixes and alternative names normally used with the cubic meter are summarized in Table 7.1. When changing prefixes, you may find it easier to use this table instead of the prefix chart on the front inside cover. Between cubic kilometer and cubic millimeter, jump the decimal point three places per prefix (or alternative name) as usual, but do *not* skip the

*The North American symbol for liter is capital L, to avoid confusion with the numeral one (1). However, the old lowercase el symbol (l) is still used in many countries.

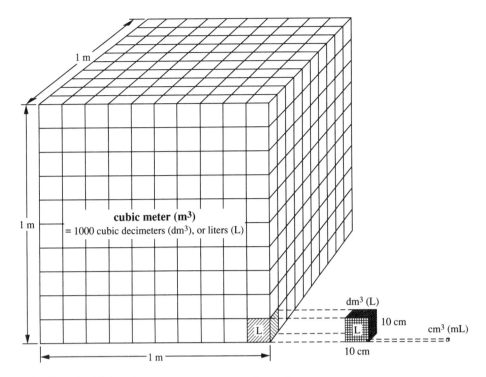

FIGURE 7.1 Multiples of cubic meter make nesting cubes.

shaded prefixes. Remember that you cannot simply substitute a prefix for its equivalent power of ten. For example, 10^{-3} m^3 is *not* a cubic millimeter (mm^3); it is a cubic decimeter (dm^3), commonly called a liter.

Volumes larger than a cubic kilometer or smaller than a cubic millimeter step by a billion (10^9) per prefix. Prefixes are impractical for such huge or tiny volumes because of the large number of placeholding zeroes that can result. Use scientific notation instead. For example, it would be better to express the volume of ice in Antarctica as $3 \cdot 10^{16}$ m^3 than as 30 000 000 km^3 (the nearest prefixed multiple).

7.5 CHANGING PREFIXES

Change prefixes by going up or down Table 7.1, jumping the decimal point three places for each multiple you pass (including no prefix, 10^0 m^3).

TABLE 7.1	Multiples of Cubic Meter		
Symbol	**Name**	**Equivalent**	**Example**
km^3	cubic kilometer	10^9 m^3	mountain
hm^3	cubic hectometer	10^6 m^3	large building
dam^3	cubic dekameter	10^3 m^3	large house
m^3	cubic meter	10^0 m^3	desk
dm^3 (L)	cubic decimeter (liter)	10^{-3} m^3	bottle
cm^3 (mL)	cubic centimeter (milliliter)	10^{-6} m^3	bean
mm^3 (μL)	cubic millimeter (microliter)	10^{-9} m^3	sand grain

- 0.002 L = 2 mL = 2 cm³
- 2 000 L = 2 m³
- 45 000 cm³ = 45 dm³ = 45 L
- 0.007 9 m³ = 7.9 dm³ = 7.9 L

To assign a prefix to a volume in scientific notation *without* referring to Table 7.1, take the *cube root* of the power of ten. Do this by dividing the exponent by 3. Then substitute a prefix for the root.

- $5.2 \cdot 10^6 \text{ m}^3 = 5.2 \cdot (10^2)^3 \text{ m}^3$
  $= 5.2 \cdot (10^2 \text{ m})^3$
  $= 5.2 \text{ hm}^3$

We substituted the prefix hecto (h) for 10^2.

- $6.92 \cdot 10^{-6} \text{ m}^3 = 6.92 \cdot (10^{-2})^3 \text{ m}^3$
  $= 6.92 \cdot (10^{-2} \text{ m})^3$
  $= 6.92 \text{ cm}^3$
  $= 6.92 \text{ mL}$

We substituted the prefix centi (c) for 10^{-2}. We may use the alternative name, milliliter (mL), for cubic centimeter (cm³).

If the exponent is not a multiple of 3, you will have to adjust the decimal point first (Section 4.7) or use the ENG display on your calculator (Section 4.8).

- $7.8 \cdot 10^7 \text{ m}^3 = 78 \cdot 10^6 \text{ m}^3 = 78 \text{ hm}^3$
- $1.46 \cdot 10^{-5} \text{ m}^3 = 14.6 \cdot 10^{-6} \text{ m}^3 = 14.6 \text{ cm}^3$

Example Problem 7.1. What is the volume of a box 20 cm wide, 34 cm long, and 15 cm deep?

Solution. The volume of a rectangular space is the product of its three length dimensions ($V = l_1 \cdot l_2 \cdot l_3$). The dimensions may have different names, such as width, height, depth, or thickness, but that is unimportant. In this problem, all dimensions have the same prefix (centi). If we ignore it, carrying it through the problem, the answer will have that prefix as well—cubic centimeter.

$$20 \text{ cm} \cdot 34 \text{ cm} \cdot 15 \text{ cm} = 10\,000 \text{ cm}^3 = 10 \text{ dm}^3 = 10 \text{ L}$$

All the data had two significant digits, so we rounded the answer to two significant digits. To simplify the answer and eliminate the placeholding zeroes, we moved the decimal point 3 places left and stepped up to the next multiple, cubic decimeter, also called a liter (Table 7.1).

We could also solve this problem by entering the prefixes on the calculator with the exponent key, E. Doing so requires twice as many keystrokes but gives an answer in cubic meters (m³), without a prefix.

20 E +/- 2 × 34 E +/- 2
× 15 E +/- 2 =

The displayed answer is 10E–3, meaning $10 \cdot 10^{-3} \text{ m}^3$, which we simplify to 10 dm³, or 10 L (Table 7.1). In this case there was no advantage in entering the prefixes.

Example Problem 7.2. What is the volume of a box 2 cm by 4 cm by 5 mm?

Solution. We cannot multiply dimensions with different prefixes, such as centimeter and millimeter. We have two choices (Section 4.8): First, we can *eliminate* the prefixes by moving decimal points or by using the exponent key on a calculator to enter the prefixes. In that case the answer will be in cubic meters (m³). Second, we can move decimal points beforehand so that all the dimensions have the *same* prefix, in which case the answer will have that prefix too. We decide that the second method is easier, and we rewrite 5 mm as 0.5 cm.

$$2 \text{ cm} \cdot 4 \text{ cm} \cdot 0.5 \text{ cm} = 4 \text{ cm}^3$$

Example Problem 7.3. What is the volume of a rectangular board 9.0 mm by 30 cm by 4.2 m?

Solution. This time we decide that it is easier to eliminate the prefixes by entering them all on the calculator. The answer will then be in cubic meters, without a prefix. We key in

9 E +/- 3 × 30 E +/- 2 × 4.2 =

The displayed answer rounds to $11 \cdot 10^{-3} \text{ m}^3$, which simplifies to 11 dm³, or 11 L (Table 7.1).

Example Problem 7.4. What is the volume of a sphere 136 cm in diameter?

Solution. The volume of a sphere is

$$V = \frac{4}{3} \pi r^3$$

We must divide the diameter by 2 to get the radius. We have two options for handling the prefix centi (10^{-2}). We can carry it along and in doing so will get an answer with the same prefix (cubic centimeter). Or we can enter the data in meters, which will give an answer in cubic meters. Choosing the first option, we enter

4 ÷ 3 × π × (136 ÷ 2) y^x 3 =

The displayed answer rounds to $1.32 \cdot 10^6 \text{ cm}^3$. To simplify this expression, by eliminating the 10^6, we step up two multiples to cubic meter, and we get 1.32 m³. The factor $\frac{4}{3}$ is exact so we rounded the answer to the three significant digits of the original data.

If we had entered the data as 1.36 m (instead of 136 cm), the answer would be in cubic meters and we would not need to simplify it.

4 ÷ 3 × π × (1.36 ÷ 2)
y^x 3 = 1.32

Example Problem 7.5. What is the volume of a shallow lake with an area of 75 km² and a mean depth of 12 m?

Solution. We can treat the lake as a thin, irregular prism. The volume of a prism is $V = Bh$, where B is the area of the base (75 km²) and h is the "height" or thickness of the prism (12 m). Remember to square the prefix kilo on square kilometer, by entering 75 km² as $75 \cdot 10^6$ m² (see Chapter 6).

$$V = Bh = (75 \cdot 10^6 \text{ m}^2) \cdot 12 \text{ m} = 900 \cdot 10^6 \text{ m}^3 = 900 \text{ hm}^3$$

Non-SI Volume Units Used in the United States

gallon	pint	bushel	cubic yard	board-foot	teaspoon	dry barrel	freight ton
quart	dry pint	cubic inch	cubic mile	acre foot	tablespoon	oil barrel	register ton
dry quart	fluid ounce	cubic foot	cord	cup	keg	liquid barrel	

Volumes of Common Figures

Rectangular solid	$V = l_1 l_2 l_3$	Prism (including cylinder)	$V = Bh$
Pyramid (including cone)	$V = \frac{1}{3} Bh$	Sphere	$V = \frac{4}{3} \pi r^3$

Problem Set 7
Volume

1. What is the quantity of space (3-dimensional) called?
2. Name the SI unit of volume.
3. What is the alternative name for cubic decimeter?
4. What is the alternative name for cubic centimeter?
5. What is the alternative name for cubic millimeter?
6. What is the alternative (formal SI) symbol for mL?
7. What is the alternative (formal SI) symbol for L?
8. What is the alternative (formal SI) symbol for μL?
9. What does the symbol cm^3 mean?
   (A) $c(m^3)$ (B) $(cm)^3$ (C) $c \cdot m^3$ (D) $c^9 \cdot m^3$

Express the following as a power of ten of the cubic meter. *Example*: cubic kilometer $(km^3) = 10^9 \ m^3$.

10. cubic centimeter (cm^3)
11. cubic hectometer (hm^3)
12. cubic dekameter (dam^3)
13. cubic millimeter (mm^3)
14. cubic decimeter (dm^3)
15. liter (L)
16. milliliter (mL)
17. microliter (μL)

Which multiple describes the volume of the following cubes?

18. A cube 100 m on a side
   (A) dm^3 (B) cm^3 (C) dam^3 (D) hm^3
19. A cube 10 cm on a side
   (A) cm^3 (B) dam^3 (C) dm^3 (D) mm^3
20. A cube 0.1 m on a side
   (A) dam^3 (B) cm^3 (C) L   (D) mm^3
21. A cube 1 cm on a side
   (A) dm^3 (B) mL   (C) L   (D) μL
22. A cube 1000 m on a side
   (A) hm^3 (B) km^3 (C) L   (D) dam^3
23. A cube 1 mm on a side
   (A) cm^3 (B) mL   (C) μL   (D) dm^3
24. A cube 10 m on a side
   (A) dam^3 (B) dm^3 (C) hm^3 (D) L

Simplify by changing prefixes (include any alternative symbols).

25. 1000 cm^3
26. 1000 m^3
27. 1 000 000 m^3
28. 0.001 m^3
29. 1000 L
30. 0.001 L
31. 53 000 L
32. 0.020 dm^3
33. 0.06 L
34. 0.025 mL
35. 0.04 m^3
36. $68 \cdot 10^9 \ m^3$
37. 0.083 cm^3
38. 71 000 hm^3
39. $36 \cdot 10^{-6} \ m^3$
40. $4.7 \cdot 10^{-3} \ m^3$

Calculate the volumes of the following objects. Answer with an appropriate prefix and correct precision. Include any alternative symbols.

41. A rectangular room 4.1 m by 8.6 m by 3.3 m
42. A rectangular gymnasium 45 m by 25 m by 8 m
43. A rectangular skyscraper 85.5 m by 91 m by 156 m
44. A box 4.5 mm by 91.4 cm by 15.3 cm
45. A cylindrical juice can 105 mm in diameter and 158 mm high
46. A rectangular reservoir 85 km by 12 km by 250 m deep
47. A rectangular board with dimensions 9.5 mm, 29 cm, and 2.2 m
48. The great pyramid of Khufu in Egypt, 147 m high with a square base 230 m on a side
49. A basketball (radius 12 cm)
50. A tennis ball (*diameter* 65 mm)
51. The moon (*diameter* 3500 km)
52. The sun (*diameter* 1.4 Gm)
53. A cinder cone (small conical volcano) 900 m in diameter and 150 m high
54. The glacial ice that covers Greenland (2.1 Mm^2 in area, mean depth 2.5 km)
55. An oxygen atom (radius = 60 pm)

Convert the following quantities to SI, using the definitions in Appendix B. Answer with an appropriate prefix and precision.

56. 350 in^3 (cubic inches)
57. A 1 gallon (U.S. liquid) jug of milk
58. 850 million board-feet of lumber
59. 13 million acre feet of water

CHAPTER 8

Mass and Density

8.1 THE KILOGRAM

Mass (*m*) is the quantity of matter. In other words, it measures "how much of something" there is. In everyday language, mass is often erroneously called "weight," a very confusing term with many different meanings (mass, force, force of gravity, density, viscosity, mass per area, etc.). In correct scientific language, weight refers to the force of gravity (Section 10.2), not mass. We should say "he has a mass of 75 kilograms," not "he weighs 75 kilograms."

The SI unit of mass is the kilogram (kg). It was originally defined in 1790 as the mass of a liter (cubic decimeter) of pure water. Because it is difficult to measure the volume of a liquid precisely, a prototype kilogram of dense, noncorroding platinum alloy was made in the nineteenth century. A kilogram is now officially defined as the mass of this object, the International Prototype Kilogram, which is carefully preserved in an underground vault at the International Bureau of Weights and Measures (BIPM*) in Sèvres, France. The kilogram is the only base unit still defined by a material artifact. The other six base units are now defined by natural phenomena that can be measured anywhere. However, the original water definition of the kilogram is close enough for ordinary purposes. The water must be at its maximum density, which is just above freezing (about 4 °C).

*From the initials of its French name, *Bureau International des Poids et Mesures.*

8.2 PREFIXES AND ALTERNATIVE NAMES

For historical reasons, the kilogram is the only *base* unit that includes a prefix (kilo). Other prefixes are attached to the word gram in the usual way, as shown in Table 8.1.

Because kilogram is the base unit, most physical laws are designed to work with kilograms, not grams. When doing calculations with these equations, ignore the prefix kilo. Use a power of ten with kilogram instead. For example, enter 9 kg as 9 kg (*not* $9 \cdot 10^3$ g); enter 5 g as $5 \cdot 10^{-3}$ kg (*not* 5 g); enter 3 Mg as $3 \cdot 10^3$ kg (*not* $3 \cdot 10^6$ g). In this case, you cannot simply substitute a prefix for the equivalent power of ten; the prefixes are all offset by one step (1000). However, in some fields, such as chemistry, equations are based on grams rather than kilograms and prefixes are used in the usual way.

A megagram (Mg = 1000 kg) is commonly called a *metric ton* in the United States or a *tonne* elsewhere. These terms are holdovers from an old version of the metric system. No prefixes should be used with them. Figure 8.1 will help acquaint you with the kilogram and its multiples.

Change prefixes in the usual way, jumping the decimal point three places per prefix (skipping the shaded prefixes on Table 1, which are not multiples of 1000).

- 2 500 mg = 2.5 g
- 5 000 kg = 5 Mg
- 0.035 kg = 35 g

	TABLE 8.1 Common Multiples of Kilogram		
Symbol	**Name**	**Equivalent**	**Approximation**
Pg	petagram (billion metric tons)	10^{12} kg	mass of 1 km^3 of water
Tg	teragram (million metric tons)	10^9 kg	mass of 1 hm^3 of water
Gg	gigagram (1000 metric tons)	10^6 kg	mass of 1 dam^3 (1000 m^3) of water
Mg	megagram (metric ton, or tonne)	10^3 kg	mass of 1 m^3 of water
kg	kilogram	*base unit*	mass of 1 dm^3 (L) of water
g	gram	10^{-3} kg	mass of 1 cm^3 (mL) of water
mg	milligram	10^{-6} kg	mass of 1 mm^3 (μL) of water
μg	microgram	10^{-9} kg	mass of 0.001 mm^3 (μL) of water

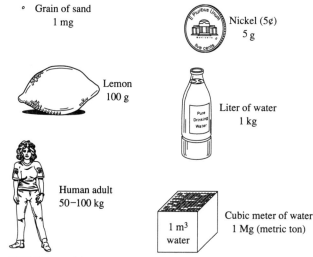

° Grain of sand
1 mg

Nickel (5¢)
5 g

Lemon
100 g

Liter of water
1 kg

Human adult
50–100 kg

Cubic meter of water
1 Mg (metric ton)

1 m³ water

FIGURE 8.1 Mass comparisons.

8.3 DENSITY

Density (D or ρ) is mass per volume.

$$D = \frac{m}{V} \quad \textit{(Definition of density)}$$

Density measures how "tightly packed" matter is. For example, lead is denser than wood. More matter occupies a given volume of lead than occupies the same volume of wood. In everyday language we might say lead is "heavier" than wood, but that is misleading since a kilogram of lead has the same mass as a kilogram of wood.

Strictly speaking, density should be measured in kilograms per cubic meter (kg/m^3)—the unit of mass divided by the unit of volume. However, for historical reasons it is often measured in kilograms per liter (kg/L) or equivalent expressions. The original metric system was designed to give water a density of exactly one kilogram per liter (1 kg/L). Water is our most important substance, and most solids and liquids have densities fairly close to that of water. We therefore find it convenient to compare densities to water's. Density is thus an exception to the rule that you should not use a prefix in the denominator of a compound unit (Section 4.5). (Remember that a liter is really a cubic decimeter, a prefixed multiple of cubic meter).

There are several equivalent density expressions that can be confusing. By definition of the liter, $kg/L = kg/dm^3$. If we multiply both numerator and denominator by 1000 (10^3), we get

$$\frac{kg \cdot 1000}{dm^3 \cdot 1000} = \frac{Mg}{m^3}$$

This expression is also called a *tonne* (or metric ton) *per cubic meter*. Note that the quantity remains unchanged, since $1000/1000 = 1$. *Dividing* both the numerator and denominator of kg/dm^3 by 1000, we get

$$\frac{kg \div 1000}{dm^3 \div 1000} = \frac{g}{cm^3} = \frac{g}{mL}$$

So the following density expressions are all equivalent and interchangeable:

$$kg/L = kg/dm^3 = g/cm^3 = g/mL = Mg/m^3$$
$$= \text{metric ton (tonne) per cubic meter}$$

In all of these compound units, the density of water equals 1. But other combinations are not necessarily the same. For example, water is 1000 kg/m^3 or 1000 g/L. These last two expressions are commonly used for gases, which have much lower densities than solids or liquids. For volumes of a cubic meter or more, prefix problems are avoided by expressing densities in mass per cubic meter. In that case, any prefix on gram is entered in the calculator (see Example Problems 8.4 and 8.5).

Example Problem 8.1. Express 1.2 g/L in mass per cubic meter.

Solution. A cubic meter is 1000 liters, so we multiply both numerator and denominator by 1000.

$$\frac{1.2 \, g \cdot 1000}{L \cdot 1000} = \frac{1.2 \, kg}{m^3}$$

Example Problem 8.2. Express 2500 kg/m^3 in mass per liter.

Solution. A liter is 1/1000 of a cubic meter, so we *divide* both the numerator and denominator by 1000.

$$\frac{2500 \, kg \div 1000}{m^3 \div 1000} = \frac{2.5 \, kg}{L}$$

Example Problem 8.3. What is the density of a 2.8 cm^3 object with a mass of 16 g?

Solution. Substitute the given values in the definition of density.

$$D = \frac{m}{V} = \frac{16 \, g}{2.8 \, cm^3} = 5.7 \, g/cm^3$$
$$= 5.7 \, kg/L = 5.7 \, Mg/m^3$$

Remember that the expressions g/cm^3, kg/L, and Mg/m^3 are equivalent and interchangeable.

Example Problem 8.4. Hoover Dam has a volume of 2.5 hm^3. If the density of concrete is 2.8 Mg/m^3, what is the mass of the concrete in the dam?

Solution. Solve the density definition for mass. For neatness we will set up the problem in fraction-bar notation (Section 2.6). Enter the prefix mega (10^6) on megagram.

$$D = \frac{m}{V}$$

$$m = DV = \frac{2.8 \cdot 10^6 \, g}{m^3} \bigg| \frac{2.5 \cdot 10^6 \, m^3}{} = 7 \cdot 10^{12} \, g = 7 \, Tg$$

The cubic meters canceled, leaving the grams of mass we were asked to find. To simplify the answer we substituted the prefix tera (T) for 10^{12}, and got teragrams (Tg). We could substitute the awkward but more common name "million metric tons" for teragram (Table 8.1).

Example Problem 8.5. If the density of air is 1.2 kg/m^3, what volume is occupied by 25 g of air?

Solution. We solve the density definition for volume. The grams (units alone) cancel, but grams and kilograms do not, so we must enter the prefix kilo (10^3) on kilogram.

$$D = \frac{m}{V}$$

$$V = \frac{m}{D} = \frac{25 \text{ g}}{} \,\bigg|\, \frac{\text{m}^3}{1.2 \text{ kg}}$$

$$= 21 \cdot 10^{-3} \text{ m}^3 = 21 \text{ dm}^3 = 21 \text{ L}$$

To solve this problem with a calculator, we enter

$$25 \;\boxed{\div}\; 1.2 \;\boxed{\text{E}}\; 3 \;\boxed{=}$$

To simplify the answer, we recalled (from Table 7.1) that 10^{-3} m^3 is a cubic decimeter, because the prefix deci is cubed ($[10^{-1}$ m$]^3 = 10^{-3}$ m^3). We can substitute the common name liter (L) for cubic decimeter (dm^3).

8.4 BUOYANCY

If an object is less dense than a surrounding fluid (gas or liquid) it is buoyant and will tend to rise or "float." Anything less than 1 kg/L will therefore float in water. Ice floats because it is less dense than water (because water has the unusual property of expanding when it freezes). On the other hand, anything denser than a surrounding fluid will sink. Rocks sink in water because their density is greater than 1 kg/L.

8.5 MIXTURES AND CONCENTRATIONS

The parts of a solution or other mixture can be expressed in various ways. If both the parts and the whole are measured as the same quantity (usually mass or volume), the units cancel and we can describe the parts as fractions of the whole. Such fractions are traditionally expressed in percent (% = 10^{-2}), per thousand (‰ = 10^{-3}), parts per million (ppm = 10^{-6}), parts per billion (ppb = 10^{-9}), and parts per trillion (ppt = 10^{-12}). For example, seawater averages 35 g of salt per kilogram of water. Since the "grams" cancel we can write this as 0.035 or 35 parts per thousand (35‰). The quantity must always be specified to prevent ambiguity. For example, air is 21% oxygen by volume, but 23% oxygen by mass.

In practice, the traditional fraction names often cause confusion; SI recommends against them. The quantity (mass or volume) is usually omitted. The symbols % and ‰ look alike and are easily confused. The abbreviation ppt is often taken to mean "parts per thousand." The words "billion" and "trillion" have different meanings in Britain and most other countries. Since water is very close to 1 kg/L, water mixtures are often expressed incorrectly with a fraction when the parts are measured by mass and the whole by volume. This is only approximately correct, and only for very dilute mixtures, where the concentration is low. For all these reasons, it is better to use SI units.

Traditional fraction	Mass per mass (SI)	Mass per volume (dilute water mixture)	
% (10^{-2})	= 10 g/kg	≈ 10 g/L	≈ 10 kg/m^3
‰ (10^{-3})	= g/kg	≈ g/L	≈ kg/m^3
ppm (10^{-6})	= mg/kg	≈ mg/L	≈ g/m^3
ppb (10^{-9})	= μg/kg	≈ μg/L	≈ mg/m^3
ppt (10^{-12})	= ng/kg	≈ ng/L	≈ μg/m^3

8.6 SUMMARY OF PREFIX COMPLICATIONS

All the prefix complications that might cause confusion have been discussed. Reviewed below are the four situations when you cannot directly substitute a prefix for its equivalent power of ten.

- *Area* (multiples of square meter). The prefix must be squared. The prefixes hecto and centi are commonly used to fill the otherwise large gap between square kilometer and square millimeter, but multiples *never* step by a thousand per prefix. A square hectometer is usually called by its alternative name, hectare. As far as prefixes are concerned, area is the most difficult quantity.

- *Volume* (multiples of cubic meter). The prefix must be cubed. The prefixes hecto, deka, deci, and centi are used to fill the otherwise large gap between cubic kilometer and cubic millimeter. The alternative names liter, milliliter, and microliter are used for the most common multiples. But multiples between kilo and milli still step by a thousand per prefix or alternative name.

- *Mass* (multiples of the kilogram). The kilogram is the only base unit that includes a prefix, so the prefix kilo may be ignored when entering the mass in most (but not all) equations and laws. Other prefixes are offset by a thousand (10^3).

- *Density* (mass per volume). Density involves the complications of both mass and volume units and is an exception to the rule that you should not use a prefix in the denominator of a compound unit. Many of the compound density multiples are equivalent.

For all other units and quantities, you may simply substitute a prefix for its equivalent power of ten, or vice versa.

Non-SI Mass Units Used in the United States

avoirdupois pound	avoirdupois dram	assay ton	point	hundredweight
avoirdupois ounce	short ton	grain	scruple	slug
troy ounce	long ton	carat	pennyweight	atomic mass unit

Non-SI Density Units Used in the United States

pound per cubic foot	pound per cubic yard	ounce per cubic inch	degrees Baumé
pound per gallon	short ton per cubic yard	pound per cubic inch	degrees Twaddell
ounce per gallon	long ton per cubic yard		

Problem Set 8
Mass and Density

1. What do you call the quantity of matter?
2. Give the two alternative names for megagram.
3. What is the original definition of a kilogram?
4. What is the modern definition of a kilogram?

Simplify the following quantities with an appropriate prefix.

5. 8 metric tons
6. 5000 g
7. 1500 mg
8. 0.054 kg
9. 15 000 kg
10. 0.02 mg
11. 0.07 g
12. 86 million kilograms
13. 34 million metric tons
14. $6.3 \cdot 10^{-3}$ kg
15. $4.5 \cdot 10^{3}$ kg
16. $1.3 \cdot 10^{-5}$ kg

Give the approximate mass of the following volumes of water.

17. 6.5 L of water
18. 12 m^3 of water
19. 30 cm^3 of water
20. 7 mL of water
21. 8.9 mm^3 of water
22. 29 dm^3 of water
23. 470 hm^3 of water
24. 31 μL of water
25. 62 km^3 of water
26. 44 dam^3 of water
27. Define density.
28. Which does *not* equal the density of water?

(A) 1 g/cm^3 (B) 1 kg/L   (C) 1 kg/m^3
(D) 1 Mg/m^3 (E) 1 g/mL   (F) 1000 g/L

29. Express 1300 kg/m^3 in mass per liter.
30. Express 3.2 g/L in mass per cubic meter.

Calculate the density of the following.

31. A 15.4 cm^3 piece of metal that has a mass of 73.6 g
32. 520 mL of a liquid that has a mass of 740 g
33. A 3.9 tonne boulder with a volume of 1.5 m^3

34. A 780 g block of wood measuring 4 cm by 9 cm by 35 cm
35. A sphere 17 cm in *diameter* that has a mass of 15 kg
36. Earth (mass = $5.98 \cdot 10^{24}$ kg, radius = 6.37 Mm)
37. Planet Saturn (mass = $5.69 \cdot 10^{26}$ kg, *diameter* = 121 Mm)
38. Would Saturn sink or float in water?
39. 9.2 g of a gas that occupies 5.5 L
40. 5.6 ng of a substance that occupies 2.0 L
41. A carbon atom (assume that it is a sphere that has a mass of 20 yg with a radius of 77 pm)

Calculate the mass of the following objects.

42. 2.3 L of mercury (density = 13.5 kg/L)
43. 35 cm^3 of iron (density = 7.9 Mg/m^3)
44. 23 hm^3 of earth (density = 1.8 Mg/m^3)

Calculate the volume of the following objects.

45. 1.00 kg of platinum (density = 21.4 kg/L)
46. 43 g of granite (density = 2.6 kg/L)
47. 15 g of air (density = 1.2 kg/m^3)

Express the following concentrations in SI units (mass per mass).

48. 35 parts per million (ppm)
49. 3%
50. 17 parts per billion (ppb)

Express the following dilute water concentrations in SI units (mass per volume).

51. 0.2%
52. 12 parts per million
53. 55 parts per trillion

Convert the following quantities to SI, using the conversion factors in Appendix B.

54. 65 ounces of copper
55. 65 ounces of gold
56. A ship cargo of 450 000 displacement tons
57. 35 000 pounds (avoirdupois)
58. Which quantity is measured with a non-SI unit called a "ton"?

(A) mass   (B) volume
(C) energy   (D) power
(E) force   (F) all of the above

CHAPTER 9

Time and Rates

9.1 THE SECOND

Thanks to an unfortunate accident of history, the SI unit of time or period (t) is the *second* (s). Prefixes are attached to second in the regular way, without complications. A day is 86.4 kiloseconds (ks) and a seasonal year (about $365\frac{1}{4}$ days) is 31.56 megaseconds (Ms). In the original metric system, the day was to be the time unit, with decimal subdivisions. But people refused to use the new units because they thought their clocks would be rendered obsolete. Clocks were expensive possessions in the late eighteenth century. So we kept the cumbersome time units invented thousands of years ago by Babylonian astrologer-priests. The second is the only SI unit older than the metric system. All of our units depend on it, and it is too late to change now.

9.2 NON-SI TIME UNITS

Days and years are natural units, but hours, minutes, weeks, and calendar months are the arbitrary inventions of ancient cultures. None are SI. However, because these traditional time units have been used worldwide for centuries, they are approved for use with SI when the second is impractical (see Table 5 on the inside back cover). But avoid them if possible because they make comparison and computation difficult. Data in traditional time units must be converted to SI (seconds) before you can use them in any SI units incorporating time, such as the watt (Chapter 12).

In geology and related sciences, which deal with very large periods, the prefixes kilo, mega, and giga are often attached to the non-SI unit *annum* (year).

$$ka = kiloannum = thousand\ years$$
$$\approx 31.6\ gigaseconds\ (Gs)$$

$$Ma = megannum = million\ years$$
$$\approx 31.6\ teraseconds\ (Ts)$$

$$Ga = gigannum = billion\ (10^9)\ years$$
$$\approx 31.6\ petaseconds\ (Ps)$$

These are not SI terms, and annum is the only non-SI time unit that takes prefixes.

Scientific calculators have a key (usually labeled $\boxed{D.MS}$ or $\boxed{H.MS}$) for converting hours, minutes, and seconds to decimal hours, or vice versa. Some models also convert to SI (seconds).

9.3 DEFINING THE SECOND

A day is 86 400 seconds:

$$\frac{60\ s}{min} \cdot \frac{60\ min}{h} \cdot \frac{24\ h}{d} = \frac{86\ 400\ s}{d}$$

However, the period of a day is very gradually increasing. Tidal friction caused by the moon's gravitational pull is slowing Earth's rotation about 50 nanoseconds per day (ns/d). Since we do not want our time unit to change, the SI second was originally defined as 1/86 400 of a mean solar day *in the year 1900*. To preserve this standard we now use an atomic clock, which does not vary and is extremely precise. The official modern definition of a second is the period of exactly 9 192 632 770 oscillations of a specified kind of radiation emitted by cesium atoms.

9.4 TIME OF DAY

The time kept by the cesium atomic clock is called *International Atomic Time* (TAI). To keep our ordinary calendar time in step with the seasons, a "leap second" is added to TAI once or twice a year. This calendar time is called *Coordinated Universal Time* (UTC). UTC is broadcast internationally on special radio frequencies and is used to set the world's clocks. UTC is essentially the same as Greenwich Mean Time (GMT), the mean solar time at the Royal Naval Observatory in Greenwich, England. GMT has been the world time standard for navigators and astronomers since the first accurate seagoing clocks were developed in the eighteenth century. UTC (or GMT) is also used in the military, where it is called "Zulu time," after the phonetic name for "Z," which stands for zero meridian (the Prime Meridian of longitude, which runs through Greenwich). In the United States you can hear UTC time signals by telephoning (303) 499-7111 or 1-900-410-TIME. UTC is reported in units of hours, minutes, and seconds from Greenwich midnight.

Example Problem 9.1. Convert 5 d 16 h 49 min 54 s to SI (seconds).

Solution. This problem illustrates the awkwardness of our traditional time units. The time we are asked to convert consists of 4 terms added together, each with different units. We refer to Table 5 for the number of seconds in a day, hour, and minute (or calculate it from their definitions). Then we convert each term to seconds and add the results. The non-SI units cancel, leaving seconds.

$$\left(5\text{ d}\cdot\frac{86\,400\text{ s}}{\text{d}}\right)+\left(16\text{ h}\cdot\frac{3600\text{ s}}{\text{h}}\right)+\left(48\text{ min}\cdot\frac{60\text{ s}}{\text{min}}\right)$$
$$+\,54\text{ s}=492\,594\text{ s}$$

9.5 SPEED AND VELOCITY

Many common measurements are *rates*. Rates consist of some quantity *per time*. For example, speed (*v*) is distance per time.

$$v=\frac{d}{t}\quad\text{(Definition of speed)}$$

The SI unit of speed is therefore a meter per second (m/s). If prefixes are needed, they should be attached to the numerator, not the denominator (write km/s, not m/ms). If possible, avoid expressing speeds with non-SI time units because they make comparison and calculation difficult. Which is faster, for example, a meter per second or a kilometer per hour? However, vehicle speeds are usually measured in kilometers per hour (km/h) because we find it convenient to compute travel times in hours, not seconds. The maximum U.S. speed limit (65 mph) is 30 m/s, or 105 km/h.

The terms *speed* and *velocity* are often used interchangeably. But in correct scientific language, velocity means "speed *in a given direction*." Velocity is a vector quantity: It has both a magnitude (measured in meters per second) and a direction.

Example Problem 9.2. The distance from Earth to the sun is 150 Gm. How much time is required for the sun's light to reach Earth?

Solution. Solve the definition of speed for time. The speed of light, *c*, rounds to 300 Mm/s (Section 5.1 or Appendix B).

$$v=\frac{d}{t}$$
$$t=\frac{d}{v}=\frac{150\text{ Gm}}{\left(\dfrac{300\text{ Mm}}{\text{s}}\right)}=150\text{ Gm}\cdot\left(\frac{\text{s}}{300\text{ Mm}}\right)$$
$$=\frac{150\cdot10^9\text{ m}\cdot\text{s}}{300\cdot10^6\text{ m}}=500\text{ s}$$

The meters cancel, leaving seconds, the unit of time. Solving the problem with a calculator is much easier than it looks above. We enter

$$150\ \boxed{\text{E}}\ 9\ \boxed{\div}\ 300\ \boxed{\text{E}}\ 6\ \boxed{=}$$

The displayed answer is 500E0, which means $500\cdot10^0$, or 500.

9.6 FLOW (VOLUMETRIC)

Flow rate, or flow, usually means volume per time. This quantity has different symbols in different fields. For illustration we will use *Q*, the symbol used by geologists for river flow, also called "discharge."

$$Q=\frac{V}{t}\quad\text{(Definition of flow)}$$

The SI unit of flow is therefore a cubic meter per second (m^3/s). If needed, prefixes or alternative names may be used in the numerator (dm^3/s or L/s, cm^3/s or mL/s, etc.). Avoid expressing flows with non-SI time units, such as L/min or m^3/h, because they make comparison and computation difficult. However, we are often interested in comparing flow rates with the volumes they produce over a day, month, or year, in which case we must convert these time units to SI (seconds).

Example Problem 9.3. In an average year, 17 km^3 of water passes down the Colorado River through the Grand Canyon. What is the average flow rate (in cubic meters per second)?

Solution. Substitute the given data in the definition of flow (above). Recall that a year (annum) is 31.6 Ms (Table 5 or Appendix B). Remember to enter the prefix mega on megasecond and to cube the prefix kilo on cubic kilometer (entering it as 10^9 m^3).

$$Q=\frac{V}{t}=\frac{17\text{ km}^3}{31.6\text{ Ms}}=\frac{17\cdot10^9\text{ m}^3}{31.6\cdot10^6\text{ s}}=540\text{ m}^3/\text{s}$$

We can calculate the flow (*Q*) through a pipe or other channel by multiplying the cross-sectional area of the channel (A_c) by the average speed of the fluid (*v*).

$$Q=A_c v\quad\text{(Flow in a channel)}$$

Example Problem 9.4. At a certain point, the cross-sectional area of a river is 350 m^2. If the mean speed of the water at that point is 3.5 m/s, what is the flow rate of the river?

Solution. Substitute the given values in the flow in a channel equation.

$$Q=A_c v$$
$$=350\text{ m}^2\cdot\frac{3.5\text{ m}}{\text{s}}$$
$$=1200\text{ m}^3/\text{s}$$

Note that square meters times meters per second equals cubic meters per second, the unit of flow.

Problem Set 9
Time and Rates

1. What is a synonym for the quantity of time?

2. What is the only SI unit that predates the metric system?

3. Which is a natural (astronomical) time unit?
   (A) week   (B) hour   (C) minute
   (D) second   (E) day

4. What is the international name and symbol for "year" (not SI)?

5. What is the original definition of the SI second?

6. Why is the period of an astronomical day increasing?

7. The fastest Olympic runners can run 800 m in 80 s. What is their average speed?

8. At freeway speed of 30 m/s, how much time is needed to travel 170 km? (Answer in both SI [seconds] and hours.)

9. How much time is required for moonlight to reach us from the moon (384 Mm away)?

10. Alpha Centauri, 40 Pm away, is the nearest star other than our sun. How much time is required for its light to reach us? (Answer in both SI [seconds] and an appropriate non-SI unit.)

11. The Amazon River has an average flow of about 200 dam^3/s, the greatest of any river in the world. At this rate, what volume of water flows down the Amazon in a year?

12. If a well pumps at a flow rate of 5 L/s, how long will it take to fill a 30 m^3 water tank? (Answer in both SI [seconds] and an appropriate non-SI time unit.)

13. If a pipe has an inside diameter of 25 mm and the flow of water through it fills a 1 L bottle in 10 s, what is the average speed of the water in the pipe?

Convert to SI (using Appendix B if needed).

14. 282 min

15. 7.83 h

16. 19.2 d

17. 3.5 years

18. 250 Ma

19. 55 mph

20. 90 km/h

21. 35 gallons per minute

22. 4.5 acre feet per day

23. 106 000 cfs

24. Convert 2 h 45 min 19 s to decimal hours.

25. Convert 3.692 h to hours, minutes, and seconds.

CHAPTER 10

Force

10.1 THE NEWTON

Force (F) is the quantity of push or pull. In his famous Second Law of Motion, Isaac Newton (1642–1727) showed that the force applied to an object equals its mass times its acceleration:

$$F = ma \qquad (Second\ Law\ of\ Motion)$$

Acceleration (a) is the *change in* velocity per time. The symbol Δ (capital delta) means "change in."

$$a = \frac{\Delta v}{t} \qquad (Definition\ of\ acceleration)$$

Since velocity (Section 9.5) is measured in meters per second (m/s), acceleration is measured in meters per second, per second. We can express this more simply as meters per second squared.

$$\text{m/s}^2 \qquad (SI\ unit\ of\ acceleration)$$

The SI unit of force is defined by the Second Law of Motion. It is therefore the unit of mass times the unit of acceleration, or a kilogram meter per second squared (kg·m/s²). For convenience this derived unit is given the special name newton (N):

$$\text{N} = \text{kg·m/s}^2 \qquad (Definition\ of\ newton)$$

In other words, a newton is the force required to accelerate a kilogram of anything by a meter per second, each second.

Earth's gravity pulls on a mass of about 102 grams—an average lemon, say—with one newton of force. We may correctly say that a lemon *weighs* about 1 N on Earth. The acceleration due to gravity on Earth, a famous constant called standard free fall or g, is about 9.8 m/s². We use g for acceleration in the Second Law of Motion in problems involving weight (the force of gravity) on or near the surface of Earth.

Example Problem 10.1. What is the weight (force of gravity) of a 65 kg woman (on Earth)?

Solution. Solve the Second Law of Motion for force; substitute the mass of the woman and the acceleration due to gravity on Earth, $g = 9.8$ m/s².

$$F = ma = 65 \text{ kg} \cdot \frac{9.8 \text{ m}}{\text{s}^2} = 640 \text{ kg·m/s}^2 = 640 \text{ N}$$

We substituted the special name newton (N) for kg·m/s².

Example Problem 10.2. What is the mass of a car that weighs 12 kN (on Earth)?

Solution. Solve the Second Law of Motion for mass; substitute the given weight (force of gravity) and the acceleration due to gravity on Earth, $g = 9.8$ m/s².

$$m = \frac{F}{a} = \frac{12 \text{ kN}}{9.8 \text{ m/s}^2} = \left(\frac{12 \cdot 10^3 \text{ kg·m}}{\text{s}^2} \right) \left(\frac{\text{s}^2}{9.8 \text{ m}} \right) = 1200 \text{ kg}$$

All the units canceled except kilogram, the unit of mass, thereby providing a check that we set up the problem correctly.

Ordinarily we would not need to work out all the units. If we are confident that we used the correct equation, we would expect the answer to be in kilograms. Recall from Chapter 8 that kilogram (not gram) is the base unit of mass, an inconsistency SI inherited from an older version of the metric system. Equations such as the Second Law of Motion therefore require a mass expressed in kilograms, not grams. However, should you forget and enter a mass in grams, you can still obtain a correct answer if you carry out all the units and pay close attention to the prefixes.

There are four known kinds of force in the universe—gravitation, the electromagnetic force, the strong (nuclear) force, and the weak (nuclear) force. All are measured in newtons.

10.2 GRAVITATION

Gravitation, or gravity, is the very weak attractive force between masses (quantities of matter). The force of gravity is also called *weight*. Force, weight, and mass are often confused because weight is the force we most often measure and because the mass of an object is usually measured indirectly, from its weight. A balance scale (Figure 10.1a) measures the mass of an object by equating its weight with the weight of known, or standard, masses. Since the local acceleration of gravity is the same on both objects, their masses are also equal.

Force is commonly measured with a spring scale. When used to measure the downward pull of gravity on an object, a spring scale measures weight (Figure 10.1b). The mass of the object can be calculated from the Second Law of Motion. For convenience, spring scales are often calibrated in mass units because the acceleration of gravity varies only slightly from place to place on Earth's surface. However, on the moon

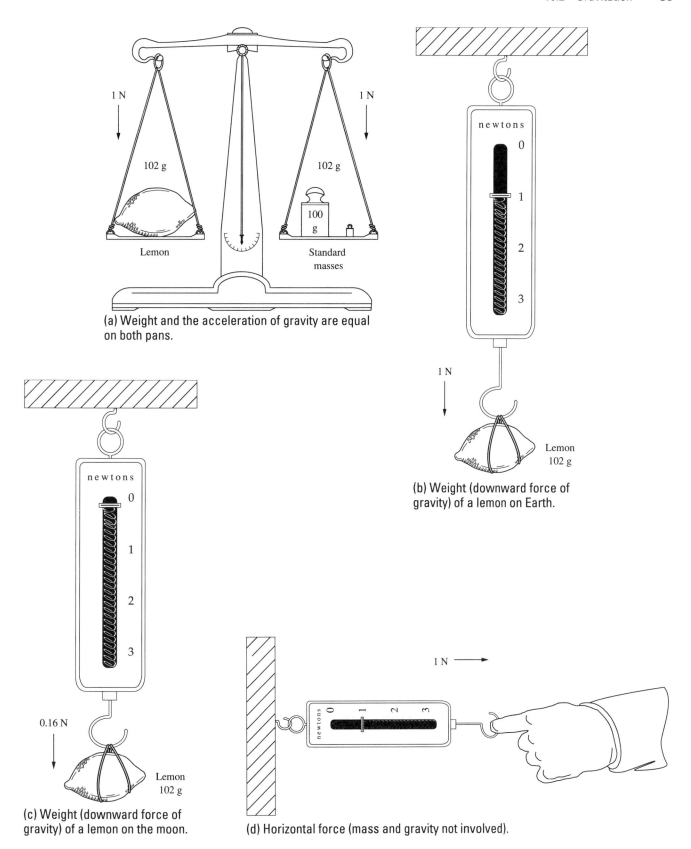

(a) Weight and the acceleration of gravity are equal on both pans.

(b) Weight (downward force of gravity) of a lemon on Earth.

(c) Weight (downward force of gravity) of a lemon on the moon.

(d) Horizontal force (mass and gravity not involved).

FIGURE 10.1   Force, weight, and mass.

(Figure 10.1c) or another body, such a scale would be useless because the acceleration of gravity is different. If a spring scale is used to measure a horizontal force (Figure 10.1d), mass and gravity are not involved and mass units are inappropriate.

In outer space the difference between mass and weight is obvious. Astronauts floating about in a spacecraft are weightless because the net gravity force pulling on them is zero. But they have not lost any of their mass (matter). If you are 75 kg on Earth, you are 75 kg everywhere.

The gravity force attracting two objects together is described by Newton's famous Law of Gravitation:

$$F_G = \frac{Gm_1m_2}{r^2} \quad \textit{(Law of Gravitation)}$$

In this equation, F_G is the force of gravitation (in newtons), m_1 and m_2 are the masses of the two objects (in kilograms), r is the distance between their centers (in meters), and G is the gravitation constant, $6.67 \cdot 10^{-11}$ N·m²/kg². Because the constant G is so small, the gravity force is only significant when at least one of the objects is a very massive body, such as a planet, moon, or star. However, gravity has an infinite range. Every object in the universe is attracted by every other object.

Example Problem 10.3. The mass of Earth is $5.98 \cdot 10^{24}$ kg, the mass of the moon is $7.35 \cdot 10^{22}$ kg, and the distance between their centers is 384 Mm. What is the force of gravitation between Earth and the moon?

Solution. Solve the Law of Gravitation for force. Enter the prefix mega (10^6) on megameter. Do not enter kilo on kilograms. (Because kilogram is the base unit, the constant G contains kilogram, not gram.)

$$
\begin{aligned}
F_G &= \frac{Gm_1m_2}{r^2} \\
&= \frac{\left(\dfrac{6.67 \cdot 10^{-11} \text{ N} \cdot \text{m}^2}{\text{kg}^2}\right)(5.98 \cdot 10^{24} \text{ kg})(7.35 \cdot 10^{22} \text{ kg})}{(384 \cdot 10^6 \text{ m})^2} \\
&= 198 \cdot 10^{18} \text{ N} \approx 200 \text{ EN}
\end{aligned}
$$

All the units canceled except newton, the unit of force. To simplify the answer we substituted the prefix exa (E) for 10^{18}. To solve this problem in a calculator, we enter

6.67 $\boxed{\text{E}}$ $\boxed{+/-}$ 11 $\boxed{\times}$ 5.98 $\boxed{\text{E}}$ 24 $\boxed{\times}$ 7.35 $\boxed{\text{E}}$ 22 $\boxed{\div}$ 384 $\boxed{\text{E}}$ 6 $\boxed{y^x}$ 2 $\boxed{=}$

10.3 ELECTROMAGNETIC FORCE

The electromagnetic force is the attractive *or* repulsive force between charges (quantities of electricity). It binds electrons and nuclei to form atoms, and binds atoms into the compounds and mixtures that make up the familiar objects of our environment. The electromagnetic force also travels through empty space as electromagnetic radiation (light, radio waves, infrared, ultraviolet, X rays, and gamma rays). Like gravity, it has an infinite range but is much stronger. See Chapters 15 and 17 for more on the electromagnetic force and electromagnetic units.

10.4 STRONG AND WEAK FORCES

The *strong* nuclear force binds protons and neutrons together to form the nucleus of an atom. It overcomes the repulsive electromagnetic force between protons that would otherwise push them apart. It is much stronger than the other forces but operates only over a very short distance—a few femtometers (1 fm = 10^{-15} m)—which is the size of a nucleus.

The *weak* nuclear force comes into play only during atomic fission (radioactive decay). It operates over an even shorter distance than the strong force—a few attometers (1 am = 10^{-18} m). The weak force is much stronger than gravitation but considerably weaker than the electromagnetic force. Experiments in the 1970s showed that the weak force is related to the electromagnetic force, both of which have been collectively called the *electroweak* force.

10.5 TORQUE

A twisting, bending, or rotational force is called torque (τ). The torque produced by the rotating crankshaft of an engine is commonly used as a measure of the engine's "strength."

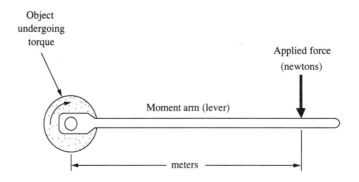

FIGURE 10.2 Torque.

Auto mechanics use torque wrenches to measure the tightness of nuts and bolts.

The force required to rotate an object depends on the point of application. For example, it takes less force to tighten a nut with a long wrench than a short one. This is the principle of the lever. Torque is thus defined as force (in newtons) times the length of a lever or "moment arm" (in meters). See

Figure 10.2. The unit of torque is therefore a newton meter ($N \cdot m$). One end of the lever (which may be real or imaginary) is at the pivot point or center of rotation. The force is applied at the other end, perpendicular to the lever. Torque gives an object angular acceleration (in rad/s^2) just as a linear force gives an object linear acceleration (in m/s^2).

Non-SI Force Units

pound-force (lbf) ("pound")

ounce-force (ozf) ("ounce")

ton-force (tonf) ("ton")

kip

kilogram-force (kgf) ("kilogram")

dyne

poundal

Non-SI Torque Units

pound-force foot ("foot pound")

pound-force inch

ounce-force inch

dyne centimeter

kilogram-force meter

Problem Set 10
Force

1. What do you call the quantity of "push" or "pull"?

2. What do you call the change in velocity per time?

3. What is the SI unit of acceleration?

4. Give Newton's Second Law of Motion in an equation.

5. Define the newton with an equation.

6. What is the correct or preferred scientific meaning of the word "weight"?

7. What is the value in SI units of g, the acceleration due to gravity on Earth, at sea level?

8. What force is required to accelerate a 25 kg object by 15 m/s^2?

9. What is the weight (force of gravity) of 1.0 m^3 of water on Earth?

10. What is the weight of a 75 kg person on Earth?

11. What is the mass of a person who weighs 780 N on Earth?

12. What force is required to lift a 1200 kg car on Earth?

13. What is the weight of a 75 kg person on the moon, where the acceleration due to gravity is 1.6 m/s^2?

14. What is the mass of a 55 kg woman on the moon?

15. What is the mass of an object that accelerates 4.2 m/s^2 when a constant force of 68 N is applied?

16. If a constant force of 958 N is applied to a 367 kg object, what is the acceleration of the object?

17. Name the force between quantities of electricity (charges).

18. Name the force of attraction between masses.

19. What is the force between nucleons (protons and neutrons) called?

20. Earth has a mass of $5.98 \cdot 10^{24}$ kg, the sun has a mass of $1.99 \cdot 10^{30}$ kg, and the distance between them is 150 Gm. What is the gravitation force between Earth and the sun?

21. What is the force of gravitation between Earth and a 75 kg person on Earth's surface? (The distance between their centers is the radius of Earth, 6.37 Mm.)

22. A bolt is tightened by applying 125 N of force to the end of a 40 cm wrench. What is the torque on the bolt?

Convert the following quantities to SI, using the definitions in Appendix B.

23. 155 lbf (pounds-force)

24. 36 kgf (kilograms-force)

CHAPTER 11

Energy

11.1 THE JOULE

Energy (E) has many names—work, heat, light, sound, potential (stored) energy, kinetic (motion) energy, radiation, infrared, ultraviolet, radio waves, X rays, gamma rays, electrical energy, chemical energy, food energy, nuclear energy, and so on. But they are all fundamentally the same. Any form of energy can change into any other form and we can measure them all with the same unit. We can define energy as the ability to do work. Work is force (F) times displacement (d)—the force applied to an object times the distance it moves as a result:

$$E = Fd \qquad \text{(Definition of energy)}$$

The SI unit of energy is therefore a newton meter (N·m)—the unit of force times the unit of distance. For convenience, this derived unit is given the special name joule (J).

$$J = N \cdot m \qquad \text{(Definition of joule)}$$

The joule (pronounced *JOOL*) was named after the English physicist James Joule (1818–1889), whose experiments showed that heat, work, and electrical energy are interchangeable.

If it takes a newton of force to keep an object moving, then a joule of energy is needed to move it one meter. For example,

it requires about 1 J of energy to lift a lemon 1 meter. Note that the force must be continually applied.

Do not confuse a newton meter of *energy* with a newton meter of *torque* (Section 10.5). Torque is an entirely different quantity and must not be measured in joules.

For many kinds of energy, the individual forces and distances are not apparent and cannot be measured directly. However, the energy can still be expressed in joules by using conversion factors determined by experiment. For example, it takes about 4.2 kilojoules (kJ) to heat a kilogram of water 1 kelvin or Celsius degree. Humans require about 8 000 to 15 000 kJ (or 8 to 15 MJ) of chemical energy per day, which we call food. A typical hamburger or serving of pie, cake, steak, chicken, or ice cream is about 1 megajoule (MJ). Lightning bolts release hundreds of megajoules. Big earthquakes and hydrogen bombs release petajoules (PJ). (See Figure 11.1.) Americans consume about a gigajoule (GJ) of commercial energy (that is, energy we pay for, like electricity and auto fuel) per person, each day. About 95% of that energy comes from nonrenewable sources—oil, natural gas, coal, and uranium. The average energy content of various fuels is listed in Appendix B under the units commonly used to measure them (see "kilogram of . . . ," "cubic meter of . . . ," "gallon of . . . ," "barrel of . . . ," "cord of . . . ," "ton of . . . ," etc.).

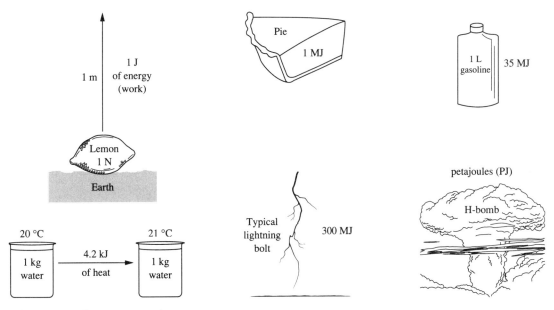

FIGURE 11.1 Energy comparisons.

When carrying out the units in equations, it is sometimes necessary to express a joule in terms of the base units. Substituting the definitions of joule and newton, we get

$$J = N \cdot m = \left(\frac{kg \cdot m}{s^2} \right) \cdot m = \frac{kg \cdot m^2}{s^2}$$

Thus, a joule may also be defined as a kilogram meter squared per second squared.

Example Problem 11.1. How much energy is required to lift a 75 kg person 4.0 m (on Earth)?

Solution. Solve the definition of energy for energy. The person is being lifted against the force of gravity, so we use the Second Law of Motion ($F = ma$) to determine the force (weight), as in Example Problem 10.1.

$$E = Fd = mad$$

$$= 75 \text{ kg} \cdot \frac{9.8 \text{ m}}{s^2} \cdot 4 \text{ m}$$

$$= 2.9 \cdot 10^3 \text{ kg} \cdot m^2/s^2$$

$$= 2.9 \text{ kJ}$$

We simplified the answer by substituting the special name joule (J) for kg·m²/s² and the prefix kilo (k) for 10^3.

11.2 MASS-ENERGY EQUIVALENCE

One could argue that energy is the most fundamental quantity in the universe, for even matter (mass) is a form of energy. Albert Einstein demonstrated this in his famous equation, where m is mass (in kilograms) and c is the speed of light (300 Mm/s):

$$E = mc^2 \quad \textit{(Einstein's equation)}$$

Example Problem 11.2. In a nuclear explosion, 85 g of matter is entirely converted to energy. How much energy is produced?

Solution. Solve Einstein's equation for energy, remembering to enter the prefix mega (10^6) on Mm/s. The equation requires mass in kilograms (not grams), so we change 85 grams to kilograms by moving the decimal point 3 places left.

$$E = mc^2$$

$$= 0.085 \text{ kg} \cdot (300 \text{ Mm/s})^2$$

$$= 7.7 \cdot 10^{15} \text{ kg} \cdot m^2/s^2$$

$$= 7.7 \text{ PJ}$$

We simplified the answer by substituting the special name joule (J) for kg·m²/s² and the prefix peta (P) for 10^{15}. To solve this problem with a calculator, we enter

.085 ☒ 300 🄴 6 $\boxed{y^x}$ 2 🟰

11.3 KINETIC ENERGY

From the definitions of energy, force, velocity, and acceleration, we can derive an equation for the energy of moving matter, called *kinetic energy:*

$$E_k = \frac{1}{2}mv^2 \quad \textit{(Kinetic energy)}$$

In this equation, E_k is the kinetic energy of the object in joules, m is its mass in kilograms, and v is its speed in meters per second. The random kinetic energy of molecules and other small particles of matter is called *heat* (Chapter 14).

Example Problem 11.3. What is the kinetic energy of a 1.5 Mg car moving at a speed of 25 m/s?

Solution. Solve the kinetic energy equation for energy. The equation requires mass in kilograms, so we enter 1.5 Mg as 1500 kg.

$$E_k = \frac{1}{2}mv^2$$

$$= 0.5 \cdot 1500 \text{ kg} \cdot (25 \text{ m/s})^2$$

$$= 470 \cdot 10^3 \text{ kg} \cdot m^2/s^2$$

$$= 470 \text{ kJ}$$

We simplified the answer by substituting the prefix kilo (k) for 10^3 and the name joule (J) for kg·m²/s².

Non-SI Energy Units Used in the United States

British thermal units (Btu) (several kinds)	erg	quad
calories (several kinds)	electron volt	Richter magnitude (earthquakes)
nutritional calorie (kilocalorie)	ton-force mile	liter atmosphere
therm	ton (nuclear)	horsepower hour
kilowatt hour	refrigeration ton	barrel of oil
foot pound-force	ton of coal equivalent	

Problem Set 11
Energy

1. Name the quantity that includes work, heat, sound, and light.

2. Give the equation used in SI to define energy.

3. Define the joule with an equation.

4. How much energy is required to lift a lemon 2 m (on Earth)?

5. How much energy is required to lift a 70 kg person 3 m (on Earth)?

6. If it takes a constant 5 kN of force to keep a car moving, how much energy is required to move it 4 km?

7. How much energy is required to lift a 210 Mg rocket 12 km (on Earth)?

8. What is the kinetic energy of a 1100 kg car moving at a speed of 30 m/s?

9. What is the kinetic energy of a 5 g bullet moving at 200 m/s?

10. If a 12 kg object has 1.4 kJ of kinetic energy, what is its speed?

11. Each second, the sun converts $4.25 \cdot 10^9$ kg of matter to energy. How much energy does the sun produce each second?

Convert the following quantities to SI, using the definitions in Appendix B.

12. 150 calories (nutritional)

13. 150 calories (thermochemical)

14. 330 kilowatt hours (of net electrical energy)

15. 11 therms (U.S.) of natural gas

16. 1200 Btu_{IT} (British thermal units, international table)

17. A 30 kiloton nuclear explosion

18. 250 foot pounds-force of work

19. 30 gallons of gasoline

20. 3700 cubic feet of natural gas (at STP)

CHAPTER **12**

Power

12.1 THE WATT

Whenever energy is moving from place to place or changing from one form to another we can refer to its power (P). This quantity is also called work rate, metabolism, heat flux, luminous flux (light flow), and luminosity (stars). Power is energy per time—the "flow rate" of energy.

$$P = \frac{E}{t} \quad \textit{(Definition of power)}$$

The SI unit of power is therefore a joule per second. For convenience this derived unit is given the special name watt (W).

$$W = \frac{J}{s} \quad \textit{(Definition of watt)}$$

The watt was named after Scottish inventor James Watt (1736–1819), who developed the steam engine, our first artificial power source.

Since a joule is the energy needed to lift a lemon 1 meter (on Earth), a watt is the power required to lift a lemon 1 meter *each second*. But let us take it back to the base units. Substituting the definitions of watt, joule, and newton, we get

$$W = \frac{J}{s} = \frac{N \cdot m}{s} = \frac{\left(\frac{kg \cdot m}{s^2}\right) \cdot m}{s} = \frac{kg \cdot m^2}{s^3}$$

So, in terms of the base units, a watt is a kilogram meter squared per second cubed. You will agree that "watt" is easier to say.

Electrical power has always been measured in watts. The "wattage" of lightbulbs and electrical appliances tells us the rate that energy flows through them. Other kinds of power should be measured in watts too. An average adult metabolizes food energy at the rate of about 150 W. Most of that power is used to keep us warm. A typical home furnace produces about 25 kW of heating power. A typical car engine produces about 100 kW of mechanical power. In 1990 the worldwide rate of consumption of commercial energy was about 10 TW (terawatts). Earth receives about 175 PW (petawatts) from the sun. (See Figure 12.1.)

When doing power calculations, it is often necessary to convert non-SI time units to SI (seconds), so it is useful to memorize the conversion factors in Table 5.

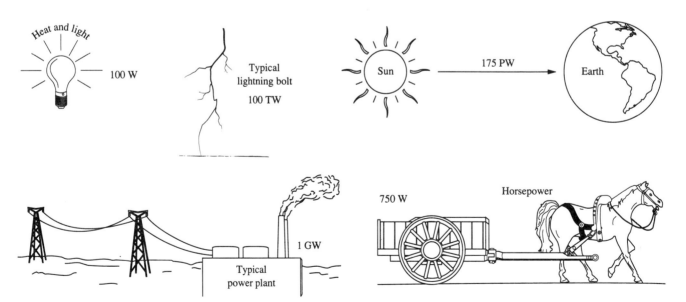

FIGURE 12.1 Power comparisons.

Example Problem 12.1. In 1990, the United States consumed about 90 EJ (exajoules) of commercial energy from all sources. What was the average consumption rate (power), in watts?

Solution. Substitute the given values in the power definition. The time t must be expressed in SI units (1 year = 31.6 Ms).

$$P = \frac{E}{t} = \frac{90 \cdot 10^{18} \text{ J}}{31.6 \cdot 10^6 \text{ s}} = 3 \cdot 10^{12} \text{ J/s} = 3 \text{ TW}$$

We substituted the prefix tera (T) for 10^{12} and the special name watt (W) for joule per second.

Example Problem 12.2. How much power is required to lift a 62 kg woman 2.5 m in 35 s (on Earth)?

Solution. Solve the power definition for power; find the energy needed to lift the woman against the force of gravity with the method of Example Problem 11.1.

$$P = \frac{E}{t} = \frac{Fd}{t} = \frac{mad}{t}$$

$$= 62 \text{ kg} \cdot \left(\frac{9.8 \text{ m}}{\text{s}^2}\right) \cdot 2.5 \text{ m} \cdot \left(\frac{1}{35 \text{ s}}\right)$$

$$= \frac{43 \text{ kg} \cdot \text{m}^2}{\text{s}^3} = 43 \text{ W}$$

To simplify the answer, we substituted the special name watt (W) for kg·m²/s³.

12.2 ENERGY EFFICIENCY

If all kinds of power are measured in watts, efficiency is readily apparent. The energy efficiency of a device equals the useable power output divided by the power input (P_o/P_i), usually expressed in percent. For example, a motor that uses 500 W of electricity to produce 250 W of mechanical power is said to be 50% efficient.

Non-SI Power Units Used in the United States

horsepower (several kinds)	foot pound-force per minute	Btu per second	Btu per hour
Btu per day	calorie per hour	ton of coal equivalent per year	absolute magnitude (stars)
calorie per second	erg per second		solar luminosities (stars)
foot pound-force per second	calorie per day	Btu per minute	
calorie per minute	quad per year	commercial refrigeration ton	

Problem Set 12
Power

1. Define power with an equation.

2. Define the watt with an equation.

3. What is the power output of an engine that can do 7.2 MJ of work in 1 min (60 s)?

4. What is the power output of a furnace that produces 150 MJ of heat per hour?

5. How much electrical energy (in joules) is used by a 100 W lightbulb burning for 1 h?

6. What is the average metabolism (power input) of a person who eats 12 MJ of food per day?

7. In 1990 Europe consumed 65 EJ of commercial energy. What was Europe's average power consumption (rate of energy use) that year, in watts?

8. How much power is required to lift a 75 kN weight to a height of 15 m in 35 s?

9. How much power is required to lift a 75 kg person (on Earth) a height of 5 m in 20 s?

10. If world petroleum reserves are 5 ZJ and petroleum is consumed at a constant rate of 4 TW, how much time will pass before the reserves are exhausted? (Answer in both SI [seconds] and an appropriate non-SI time unit.)

11. A river (on Earth) flows at a rate of 150 m³/s and falls over a 35 m waterfall. What is the power of the falling water?

12. What is the efficiency (in percent) of a motor that uses 600 W of electrical power to produce 350 W of mechanical power?

13. A furnace consumes 55 m³ (at STP) of natural gas per day. What is its power input (energy consumption rate) in watts? (See Appendix B.)

Convert the following to SI, using the definitions in Appendix B.

14. 140 horsepower (mechanical)

15. 2100 calories (nutritional) per day

16. 85 quads per year

17. 890 foot pounds-force per second

CHAPTER 13

Pressure and Stress

13.1 THE PASCAL

Pressure (*p*) is force per area, a quantity also called *stress*.

$$p = \frac{F}{A} \quad \textit{(Definition of pressure)}$$

The SI unit of pressure is therefore a newton per square meter (N/m^2). For convenience this derived unit is given the special name pascal (Pa).

$$Pa = N/m^2 \quad \textit{(Definition of pascal)}$$

The pascal (pronounced *PASS-kuhl*) was named after French mathematician Blaise Pascal (*pas-KAHL*) (1623–1662), who first explained pressure in fluids (Pascal's Principle). Remember that, regardless of the way the person's name is pronounced, SI unit names are stressed on the first syllable.

Recall that a newton is about the weight of a lemon on Earth. So a lemon resting on a meter-square tabletop is exerting a pascal of pressure, spread over the whole tabletop. That's not much. You can barely feel 1 Pa of pressure on your skin. But healthy human ears can hear sound waves with a pressure amplitude of as little as 20 µPa (micropascals).

At sea level, atmospheric pressure is about 101 kPa (pronounced *KILL-oh-pass-kuhlz*). It varies a few kilopascals with the weather. Imagine 101 000 lemons piled up on a meter-square surface. Atmospheric pressure is measured with a barometer. It drops rapidly with elevation because the mass, and therefore the weight, of air decreases aloft. The pressure on top of Mt. Everest (9 km above sea level) is only about 30 kPa. Deep in the ocean, water pressure is measured in megapascals (MPa). The pressure at the center of Earth is estimated at 300 GPa (gigapascals).

We can express the pascal in terms of the base units by substituting the definition of the newton:

$$Pa = \frac{N}{m^2} = \frac{\left(\dfrac{kg \cdot m}{s^2}\right)}{m^2} = \frac{kg \cdot m}{s^2 \cdot m^2} = \frac{kg}{s^2 \cdot m}$$

Thus a pascal equals a kilogram per second squared meter.

Example Problem 13.1. If a man weighs 840 N and the soles of his shoes have a total area of 780 cm^2, what pressure does he exert on the floor, when standing?

Solution. Substitute the given values in the definition of pressure. Remember to square the prefix centi on square centimeter, entering the area as $780 \cdot 10^{-4}$ m^2. Alternatively, we can move the decimal point four places left and enter the area as 0.078 m^2 (see Table 6.1).

$$P = \frac{F}{A} = \frac{840 \text{ N}}{780 \cdot 10^{-4} \text{ m}^2} = 11 \cdot 10^3 \text{ N/m}^2 = 11 \text{ kPa}$$

To simplify the answer we substituted the prefix kilo (k) for 10^3 and the special name pascal (Pa) for N/m^2.

13.2 GAGE VERSUS ABSOLUTE

The examples in the previous section are *absolute* pressures. Absolute pressure approaches zero in interstellar space, where almost no matter is present. Thus space approaches an absolute vacuum (absolute emptiness). An absolute pressure can never be negative. However, for everyday purposes on Earth we often refer to the pressure *greater than atmospheric*. We call it *gage* pressure because it is the quantity measured on an ordinary pressure gage. Gage pressure equals absolute minus atmospheric.

$$p_{\text{gage}} = p_{\text{abs}} - p_{\text{atmos}} \quad \textit{(Definition of gage pressure)}$$

Gage pressure may be negative if absolute is less than atmospheric, in which case it is called a *relative vacuum*. Vacuum pumps and vacuum cleaners create relative vacuums. If there is a chance of confusion, you should specify whether a pressure measurement is gage or absolute. Automobile tires are inflated to a gage pressure of 200 to 300 kPa. It takes several kilopascals (gage) to inflate a toy balloon. Human blood pressure is about 7 to 20 kPa (gage).

Example Problem 13.2. If the air in a tire has a gage pressure of 245 kPa and the atmospheric pressure is 95 kPa, what is the absolute pressure in the tire?

Solution. Solve the definition of gage pressure for absolute pressure, and substitute the given values.

$$p_{\text{gage}} = p_{\text{abs}} - p_{\text{atmos}}$$

$$p_{\text{abs}} = p_{\text{gage}} + p_{\text{atmos}} = 245 \text{ kPa} + 95 \text{ kPa} = 340 \text{ kPa}$$

Non-SI Pressure Units Used in the United States

pound-force per square inch (psi)

kilobar

pound-force per square foot (psf)

millimeter of mercury (blood pressure)

decibel (sound pressure)

ton-force per square inch

inch of mercury (weather)

kip per square inch (ksi)

inch of water

ounce-force per square inch

foot of air

standard atmosphere

millibar (weather)

technical atmosphere

bar

dyne per square centimeter

kilogram-force per square meter

kilogram-force per square centimeter

Problem Set 13
Pressure and Stress

1. Define pressure with an equation.

2. Define the pascal with an equation.

3. What is the mean atmospheric pressure at sea level?

4. A flat roof has an area of 120 m^2 and is covered with a uniform layer of snow. If the snow weighs 54 kN, what pressure does it exert on the roof?

5. A 900 kg water bed covers 3 m^2 of floor area. What pressure does the bed exert on the floor?

6. The hooves of a 600 kg cow have a total area of 800 cm^2. What pressure does the cow exert standing on the ground?

7. The hydrogen gas in a cylinder has a pressure of 16 MPa. What force does it exert on 1 cm^2 of the cylinder wall?

8. The air in an automobile tire has a *gage* pressure of 250 kPa at sea level. What is the *absolute* air pressure in the tire?

9. What is the approximate total weight (force of gravity) of Earth's atmosphere? (The radius of Earth is 6.37 Mm. The surface area of a sphere is $A = 4\pi r^2$.)

Convert the following to SI, using the definitions in Appendix B.

10. An atmospheric pressure of 985 millibars

11. A pressure of 5 atmospheres (standard)

12. Blood pressure of 150 mm of mercury (torrs)

13. Tire pressure of 35 psi (gage)

14. The pressure amplitude of a 90 decibel noise (C-range)

15. 450 kgf/cm^2 (kilograms-force per square centimeter)

CHAPTER 14

Temperature and Heat

14.1 HEAT

The tiny particles that make up matter (molecules, atoms, ions, etc.) are in constant motion. They have kinetic energy, moving randomly in different directions at different speeds. We call this kind of internal energy *heat*. Heat can flow through matter as particles collide with adjacent particles; this process is called *conduction*. Like all forms of energy, heat is measured in joules. However, the total heat energy of an object is very difficult to measure. Normally, we must be content with the difference in heat or heat flow between two objects.

A moving piece of matter, such as a falling rock, also has ordered or *external* kinetic energy (Section 11.3). This is not the same as heat because it is not random—all the particles are moving together. But any form of energy may become heat, or vice versa. Energy cannot be created or destroyed, but in a closed system (a perfectly insulated space, where no energy flows in or out) it becomes increasingly disordered as "waste heat." This principle is known as the Second Law of Thermodynamics.

14.2 TEMPERATURE: THE KELVIN

Temperature (T) is the *average* heat energy per particle. In other words, it is the mean random kinetic energy *per* molecule, atom, or ion. We do not sense this motion directly because the particles are usually less than a nanometer (1 nm = 10^{-9} m), but we can feel it as temperature. The hotter something is, the more vigorously its particles are moving.

Temperature is not measured in joules per particle, as you might expect. The kinetic energy of the particles is all but impossible to measure directly. The SI unit of temperature is the kelvin (K). It was named after British physicist William Thomson (1824–1907), who held the aristocratic title of Baron Kelvin.

Zero kelvin is the coldest theoretical temperature, called *absolute zero*. Nothing can be quite that cold, although we have come very close in the laboratory. Since nothing can get below absolute zero, there are no negative temperatures on the kelvin scale. That is why it was adopted for SI.

Temperature should not be confused with the quantity of heat. Measured in kelvins, temperature is the *average* random kinetic energy *per particle*. Heat, measured in joules, is the *total* random kinetic energy of all the particles in something. For example, a cup of hot coffee at 320 K is hotter than a bathtub full of water at 310 K. The coffee has more energy per molecule (temperature), but the bathwater has more heat because it has so many more molecules. The heat energy in an object is very difficult to measure. Temperature is easy to measure with a thermometer or other device.

14.3 DEGREES CELSIUS

For everyday temperatures, SI also employs the old metric temperature scale, degrees Celsius* (°C), invented by Swedish astronomer Anders Celsius in 1742. Celsius degrees are exactly the same size as kelvins but the zero point is shifted to the temperature of freezing water, 0 °C, or 273 K. At sea-level pressure (101 kPa), water boils at 100 °C, or 373 K. Temperatures below water's freezing point are negative numbers on the Celsius scale. That does not make much sense, because the particles do not have "negative energy." For most energy calculations, a temperature in degrees Celsius must be converted to kelvins. To convert from degrees Celsius to kelvins, add 273. To convert from kelvins to degrees Celsius, subtract 273 (the exact conversion is 273.15).

14.4 KELVIN VERSUS DEGREE CELSIUS

When referring to a temperature *difference* or interval, a Celsius degree is identical to a kelvin. In that case, degree Celsius is an alternative name for the kelvin and may be used interchangeably. For example, the units joule per kelvin (J/K) and joule per degree Celsius (J/°C) are identical, although joule per kelvin is preferred. However, when referring to *the* temperature of an object—the more common situation—the two units are *not* the same and must be converted by adding or subtracting 273.15. At extremely high temperatures the difference between the two scales becomes insignificant and can be ignored. For example, 1 000 000 °C equals 1 000 273 K but you can round it to 1 MK because such high temperatures cannot be measured precisely.

The use of degree Celsius is an unfortunate complication that became part of the International System because it was so widespread outside the United States. A quick way to get comfortable with kelvins and degrees Celsius is to learn a few reference temperatures like those in Figure 14.1.

*Formerly called *degrees centigrade*.

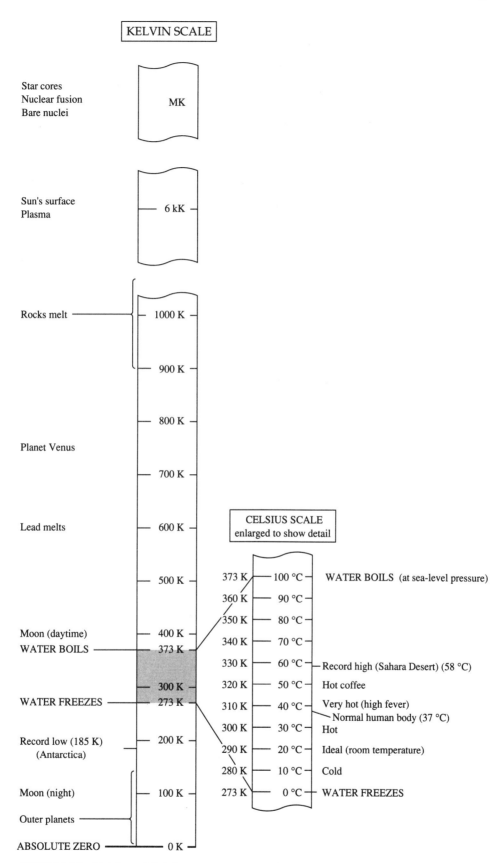

FIGURE 14.1   Kelvin and Celsius temperature scales.

14.5 HEAT FLOW

In solving the example problems in this and following sections, we will not use any prescribed formulas or variable forms of equations (with italic letters). Instead, a useful technique called "dimensional analysis" is employed. We solve the problem by manipulating the units algebraically.

Like all types of power, or energy flow, heat flow is measured in watts. The flow of heat or other energy through a surface or area is therefore measured in watts per square meter (W/m^2). This quantity has many names such as intensity, irradiance, power density, and heat flux density, and is represented with various symbols. But they all measure power per area.

Example Problem 14.1. A solar collector on the roof of a building has an area of 25 m^2. If it is aimed directly at the sun and if the sun's intensity is 1200 W/m^2, how much solar power (in watts) falls on the collector?

Solution. No special laws are involved in this problem. We are asked to find the number of watts falling on a certain number of square meters, given that a certain number of watts falls on each square meter. In problems of this sort, we can simply multiply or divide the data so that all the units cancel except the one we were asked to find (watts in this case). For neatness, we use the fraction bar notation described in Section 2.6.

$$\frac{25 \text{ m}^2 \quad \left| \quad 1200 \text{ W}\right.}{\text{m}^2} = 30 \cdot 10^3 \text{ W} = 30 \text{ kW}$$

The square meters canceled, and we were left with watts. There is no other way we could have manipulated the data to end up with watts. We substituted the prefix kilo (k) for 10^3.

14.6 THERMAL CONDUCTANCE

Heat flows from a hotter to a colder object. Often we are interested in the heat flow resulting from (or *per*) a given temperature difference. For example, we may wish to measure the heat flow through the walls of a building. This quantity has various names such as thermal conductance or coefficient of heat transfer. The unit is therefore a watt per square meter, per kelvin.

$$\frac{\left(\dfrac{W}{m^2}\right)}{K}$$

We can express this more simply; we eliminate the fraction within a fraction by inverting the denominator (kelvin) and multiplying to get

$$\frac{W}{m^2 \cdot K}$$

This unit is pronounced "watt per square meter kelvin." We can write this on one line as $W/(m^2 \cdot K)$ or $W \cdot m^{-2} \cdot K^{-1}$.

However, $W/m^2/K$ would be incorrect because it is ambiguous. It could mean either $(W/m^2)/K$, the same as the correct unit, or $W/(m^2/K) = W \cdot K/m^2$, a completely different expression. In compound quantities like this with complex division (fractions within fractions), parentheses or negative exponents must be used to prevent ambiguity.

14.7 THERMAL INSULANCE (INSULATION VALUE)

The reciprocal of thermal conductance has been given various names such as thermal insulance and insulation value. Regardless of its name, the unit is the reciprocal of the unit described in the last section, which is $K \cdot m^2/W$, pronounced kelvin square meter per watt. This quantity is commonly used to compare the heat insulating values of different building materials. (In the United States it is measured with a non-SI unit called an "R-value." See Appendix B for the conversion factor.) Other related heat flow quantities are listed in Chapter 20.

Example Problem 14.2. The roof and outside walls of a house have a total area of 300 m^2. If it takes 2000 W of heating power to keep the inside of the house 20 °C warmer than the outside air, what is the average insulating value of the walls and roof (in $K \cdot m^2/W$)? (For simplicity, ignore any solar energy input or heat flow between the house and the ground).

Solution. We are given a temperature *difference*, not *the* temperature, so we can rewrite 20 °C as 20 K. Because no special laws are involved, we simply multiply or divide the data to yield an answer with the unit we want ($K \cdot m^2/W$).

$$\frac{20 \text{ K} \quad \left| \quad 300 \text{ m}^2\right.}{2000 \text{ W}} = 3 \text{ K} \cdot m^2/W$$

There is no other way we could have manipulated the data algebraically to yield an answer in kelvin square meters per watt.

Example Problem 14.3. The external surface of a building has an average thermal insulance of 4.0 $K \cdot m^2/W$ and a total area of 650 m^2. How much heating power (in watts) is needed to keep the inside of the building 15 °C (15 K) warmer than the outside?

Solution. Again we arrange the data so that all the units cancel except the one we were asked to find, watts. To achieve this, we must divide by the insulating value (invert and multiply).

$$\frac{15 \text{ K} \quad \left| \quad 650 \text{ m}^2 \quad \right| \quad W}{4.0 \text{ K} \cdot m^2} = 2400 \text{ W} = 2.4 \text{ kW}$$

There is no other way we could have manipulated the data algebraically to give an answer in watts.

14.8 SPECIFIC HEAT

The energy needed to change the temperature of a substance is called its specific heat capacity, or specific heat. This quantity tells us how many joules of energy are needed to raise the temperature of a kilogram of the substance by 1 kelvin (or Celsius degree). The unit is therefore

$$\frac{\left(\dfrac{J}{kg}\right)}{K}$$

We can write it more simply as

$$\frac{J}{kg \cdot K}$$

We may write this unit, pronounced joule per kilogram kelvin, on one line as J/(kg·K) or J·kg^{-1}·K^{-1} (but *not* J/kg/K). The specific heat of various substances is determined experimentally. It varies with the energy state (gas, liquid, or solid). The specific heat of liquid water is unusually high, about 4.2 kJ/(kg·K).

Example Problem 14.4. How much energy is needed to raise the temperature of 20 L of water by 15 °C?

Solution. Recall from Chapter 8 that a liter of water has a mass of 1 kg, so we can express 20 L of water as 20 kg. We are given a temperature difference (not *the* temperature), and may therefore substitute 15 K for 15 °C. Since no special laws are involved, we simply multiply the data by the specific heat of water so that all the units cancel except the one that goes with the quantity we were asked to find (joule). Prefix switching is avoided by entering the prefix kilo (10^3) of kilojoules on the calculator.

$$\frac{4.2 \text{ kJ}}{kg \cdot K} \cdot \frac{20 \text{ kg}}{} \cdot \frac{15 \text{ K}}{} = 1.3 \cdot 10^6 \text{ J} = 1.3 \text{ MJ}$$

To simplify the answer we substituted the prefix mega (M) for 10^6.

14.9 LATENT HEAT

Matter can exist in four common energy states, depending on its temperature and pressure. Solid, liquid, and gas are the most familiar states on Earth. Plasma (ionized gas) is the most common state elsewhere in the universe. The sun and other stars are plasma.

When a substance is raised to a higher state—melts, vaporizes, or ionizes—it absorbs energy. When it drops to a lower state—freezes or condenses—it releases energy. This energy is called *latent* ("hidden") *heat* because it cannot be sensed or detected as a change in temperature. For example, when water reaches its boiling point, its temperature remains the same even though it continues to be heated. Latent heat is the energy used in breaking the electromagnetic bonds that hold atoms together in solids and liquids. It is usually measured in joules per kilogram (J/kg). The latent heat of vaporization for water is quite high—2.3 MJ/kg—for it takes a great deal of energy to evaporate water. Temperature is not involved in latent heat calculations.

Example Problem 14.5. How much energy is required to completely evaporate (vaporize) a shallow lake with a surface area of 3.6 km^2 and an average depth of 1.8 m?

Solution. We are asked for energy (in joules). The problem involves the latent heat of vaporization of water, which is 2.3 megajoules per kilogram (above). We are given the dimensions of the lake in meters and square meters, which, when multiplied together, yield cubic meters (the unit of volume). All the units must cancel except joules, so we must introduce a factor that relates kilograms of water to cubic meters of water (to cancel the kilograms and cubic meters). That factor is the density of water, 1 Mg/m^3 or 1000 kg/m^3 (Section 8.3). We can treat the lake as an irregular prism with a base area of 3.6 km^2 and a "height" of 1.8 m (see Example Problem 7.5). We set up the data so that all the units cancel except joules. We multiply base area times depth (to give volume) times the density of water (to give mass) times the latent heat of water (to give energy). Remember to enter the prefix mega (10^6) of megajoule and to square the prefix kilo of square kilometer (enter 3.6 km^2 as 3.6 · 10^6 m^2).

$$\frac{3.6 \cdot 10^6 \text{ m}^2}{} \cdot \frac{1.8 \text{ m}}{} \cdot \frac{1000 \text{ kg}}{\text{m}^3} \cdot \frac{2.3 \text{ MJ}}{\text{kg}}$$

$$= 15 \cdot 10^{15} \text{ J} = 15 \text{ PJ}$$

To simplify the answer, we substituted the prefix peta (P) for 10^{15}. Note that 15 petajoules is an enormous amount of energy, similar to that released by a nuclear bomb.

14.10 TEMPERATURE IN IDEAL GASES

In an ideal (perfect) gas, the molecules are completely free to move about, bouncing off each other with perfect elasticity. They are not influenced by electromagnetic forces between them. There are no ideal gases, but real gases at reasonably low density come close—air, for example. For an ideal gas, we can relate temperature and the kinetic energy of the particles with a simple equation.

$$E_k = \frac{3kT}{2} \quad \textit{(Kinetic energy of an ideal gas molecule)}$$

E_k is the random kinetic energy (heat energy) of an average molecule in joules, T is the temperature of the gas in kelvins, and k is Boltzmann's constant ($\approx 1.380\ 658 \cdot 10^{-23}$ J/K). The constant k, determined experimentally, is in effect a "conversion factor" between the kelvin and joule. We can use this equation to find the kinetic energy of an average molecule of ideal gas. Then we can substitute that energy and the mass of a molecule of the gas into the equation for kinetic energy (Section 11.3) to find the speed of an average molecule. (Chapter 18 explains how to find the molecular mass of a substance.)

Example Problem 14.6. A molecule of hydrogen, H_2, has a mass of $3.35 \cdot 10^{-27}$ kg. What is the mean velocity of hydrogen molecules in an ideal gas at a temperature of 300 K?

Solution. First solve the equation in Section 14.10 for energy.

$$E_k = \frac{3kT}{2}$$

$$= \frac{3}{2} \left| \frac{1.38 \cdot 10^{-23} \text{ J}}{\text{K}} \right| 300 \text{ K}$$

$$= 6.21 \cdot 10^{-21} \text{ J} = 6.21 \text{ zJ}$$

To simplify the answer, we substituted the prefix zepto (z) for 10^{-21}. Now we can solve the kinetic energy equation for speed, substituting the energy value just obtained and the given mass of a hydrogen molecule.

$$E_k = \tfrac{1}{2} m v^2$$

$$v = \sqrt{\frac{2E_k}{m}} = \sqrt{\frac{2(6.21 \cdot 10^{-21} \text{ J})}{3.35 \cdot 10^{-27} \text{ kg}}} = 1930 \sqrt{\frac{\text{J}}{\text{kg}}} = 1930 \text{ m/s}$$

It is not obvious that the unit $\sqrt{\text{J/kg}}$ we obtained equals m/s. If we have used the correct equations and have done the mathematics properly, we can be confident that the answer is in meters per second, because that is the unit of velocity. But let us see if this is true. Recall from Section 11.1 that, in terms of the base units, $J = kg \cdot m^2/s^2$. Substituting this definition we get

$$\sqrt{\frac{\text{J}}{\text{kg}}} = \sqrt{\frac{\frac{\text{kg} \cdot \text{m}^2}{\text{s}^2}}{\text{kg}}} = \sqrt{\frac{\text{m}^2}{\text{s}^2}} = \frac{\text{m}}{\text{s}}$$

The square root of joules per kilogram is indeed meters per second.

In solids and liquids the concept of temperature is more complicated because the particles are very close together. Electromagnetic forces between the particles prevent them from moving freely. The ideal gas laws do not apply.

14.11 DEFINING THE KELVIN

You should not think of absolute zero as a state in which all molecular motion ceases. As absolute zero is approached, the kinetic energy of the molecules approaches a definite minimum, but not zero. The kelvin scale is formally defined without reference to energy or joules. It is based on the law of physics that states that the pressure of an ideal gas varies directly with its temperature, if volume is held constant. Absolute zero (0 kelvin) is the temperature at which such a gas would have no pressure.

Because absolute zero is an unreachable standard, the kelvin scale is officially defined by the temperature of the triple point of water. In order that a kelvin would be the same size as the older Celsius degree, this reference temperature was arbitrarily defined as 273.16 K. The triple point of a substance is the state in which gas, liquid, and solid phases are all in equilibrium. We do not ordinarily experience this condition with water because it occurs at a very low pressure (611 Pa). But water's triple point is very close to its ordinary freezing point of 273.15 K (0 °C).

The official definition of the kelvin just described can be applied with devices such as a constant-volume gas thermometer. But these methods are very difficult and time-consuming to use. In practice, precise temperatures are measured by comparison with a set of fixed points and standard thermometers specified by the CGPM (General Conference on Weights and Measures) that are called the "International Practical Temperature Scale" (IPTS).

Non-SI Temperature Units Used in the United States

Degrees Fahrenheit   Degrees Rankine

Problem Set 14
Temperature and Heat

1. What do you call the total random kinetic energy of the molecules (etc.) in a sample of matter?

2. What do you call the mean random kinetic energy per molecule (etc.)?

3. What is the SI unit of temperature?

4. Name the old metric unit of temperature (used with SI).

5. Name the SI unit of heat.

6. What is the coldest theoretical temperature called?

7. What is the advantage of kelvins over degrees Celsius?

8. Convert 27 °C to kelvins.

9. Convert −173 °C to kelvins.

10. Convert 323 K to degrees Celsius.

11. Convert 15 000 000 °C to kelvins (two significant digits).

Give the temperature of the following in both kelvins and degrees Celsius (to the nearest whole number).

12. Absolute zero

13. Freezing water

14. Boiling water (at 101 kPa)

15. Ideal (room) temperature

16. Normal human body temperature

17. The sun's surface (approximate)

18. Record high air temperature (Sahara Desert)

19. Record low air temperature (Antarctica)

20. Convert 104 °F to kelvins and degrees Celsius (see Appendix B).

21. Which expression is correctly written?

    (A) watts per square meter per kelvin

    (B) $W/m^2/K$

    (C) $W/(m^2 \cdot K)$

22. On a certain day, the intensity of the sun's rays is 1300 W/m^2. How much solar power (in watts) falls on a 0.31 m^2 collector aimed directly at the sun?

23. The external surfaces of a building have a total area of 250 m^2 and a thermal insulance (insulation value) of 3.5 $K \cdot m^2/W$. What heating power output (in watts) is required to maintain an inside temperature 25 °C warmer than the outside temperature?

24. A swimming pool holds 250 m^3 of water. How much energy (in joules) is required to raise the temperature of the water 10 °C?

25. How much energy is required to completely evaporate a pond that has a surface area of 25 ha (hm^2) and an average depth of 1.2 m?

26. Name the condition in which gas, liquid, and solid phases are all in equilibrium.

27. The official definition of the kelvin is based on what reference temperature?

28. What is the temperature (in kelvins) of an ideal gas in which the average molecule has 7.2 zJ of random kinetic (heat) energy?

29. What is the average random kinetic (heat) energy of a molecule of ideal gas at 235 K?

30. An oxygen molecule (O_2) has a mass of $53 \cdot 10^{-27}$ kg. What is the average speed of oxygen molecules at 320 K (assuming ideal gas conditions)?

CHAPTER 15
Frequency

15.1 THE HERTZ

In the physical sciences, frequency means *per time*—the reciprocal or inverse of time. It measures how frequently a periodic phenomenon occurs. The symbol for frequency is f or v (the lowercase Greek letter *nu*).

$$f = \frac{1}{t} \quad \text{(Definition of frequency)}$$

The SI unit of frequency is therefore 1 per second. For convenience, this derived unit is given the special name hertz (Hz).

$$\text{Hz} = 1/\text{s} \quad \text{(Definition of hertz)}$$

The hertz was named after German physicist Heinrich Hertz (1857–1894), who discovered radio waves.

Frequency and period (or time) are actually two different ways of expressing the same quantity. We sometimes find it more convenient to measure frequency, especially for phenomena that travel as waves, such as sound and electromagnetic radiation. For all kinds of waves, there is a simple relationship that follows from the definitions of velocity and frequency:

$$v = \lambda f \quad \text{(Wave law)}$$

In this equation, v is the velocity of the wave in meters per second, λ is the wavelength in meters, and f is the frequency in hertz (1/s). The symbol λ is the lowercase Greek letter *lambda*.

The healthy human ear can detect vibrations (waves in matter) between 20 Hz and 20 kHz, which we call *sound*. Thunder is about 200 Hz. The frequency of sounds is commonly called *pitch*.

Example Problem 15.1. If the speed of sound is 330 m/s, what is the wavelength of sound with a frequency of 3.4 kHz?

Solution. Solve the wave law for frequency and substitute the given values. Substitute the definition 1/s for Hz, and enter the prefix kilo (10^{-3}).

$$v = \lambda f$$

$$\lambda = \frac{v}{f} = \frac{\left(\dfrac{330 \text{ m}}{\text{s}}\right)}{3.4 \text{ kHz}} = \frac{330 \text{ m}}{\text{s}} \cdot \frac{\text{s}}{3.4 \cdot 10^3}$$

$$= 97 \cdot 10^{-3} \text{ m} = 97 \text{ mm}$$

The seconds canceled, leaving meter, the unit of wavelength. We substituted the prefix milli (m) for 10^{-3}.

15.2 ELECTROMAGNETIC RADIATION

Electromagnetic radiation is the form of energy produced whenever charges (quantities of electricity) are accelerated. Matter is made of charged particles (electrons and protons) (Section 17.1). Since these particles are in constant motion, accelerating back and forth, all matter emits radiation. Radiation travels through a vacuum (empty space) at exactly 299 792 458 m/s, the constant known as the speed of light (c). Substituting c for the velocity in the wave law we get

$$c = \lambda f \quad \text{(Wave law for radiation in vacuum)}$$

Because we detect them with different means, different frequencies (or wavelengths) of radiation have different traditional names—radio waves, light, infrared, ultraviolet, X rays, and gamma rays (see Figure 15.1). But all are fundamentally the same and obey the same laws.

Example Problem 15.2. What is the frequency of 250 nm ultraviolet radiation?

Solution. Solve the wave law for radiation for frequency.

$$c = \lambda f$$

$$f = \frac{c}{\lambda} = \frac{300 \text{ Mm/s}}{250 \text{ nm}} = \frac{300 \cdot 10^6 \text{ m}}{\text{s}} \cdot \frac{1}{250 \cdot 10^{-9} \text{ m}}$$

$$= \frac{1.2 \cdot 10^{15}}{\text{s}} = 1.2 \text{ PHz}$$

To simplify the answer, we substituted the prefix peta (P) for 10^{15} and the special name hertz (Hz) for 1/s.

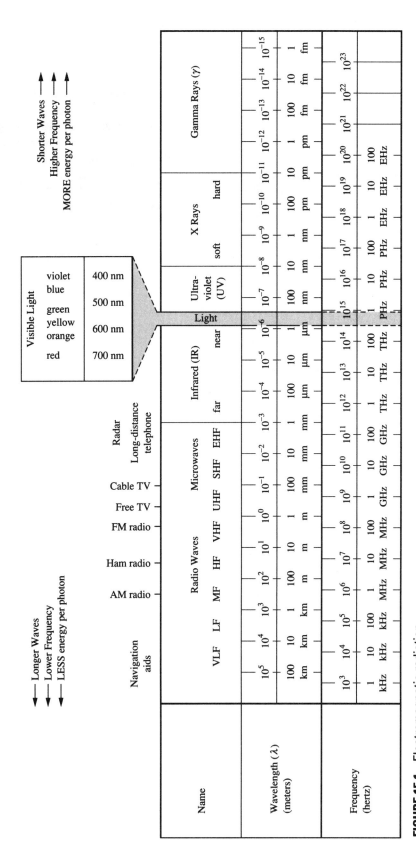

FIGURE 15.1 Electromagnetic radiation.

In some respects radiation behaves as a stream of particles, not a wave. Particles of radiation are called photons and obey Planck's Law:

$$E_p = hf \quad \text{(Planck's Law)}$$

E_p is the energy of a photon in joules, f is the frequency in hertz, and h is Planck's constant ($\approx 6.626 \cdot 10^{-34}$ J·s).

Example Problem 15.3. What is the energy of a photon of blue light with a wavelength of 450 nm?

Solution. Substitute the given value in Planck's Law, using the wave law to convert wavelength to frequency. Be sure to enter all prefixes.

$$E_p = hf = h \cdot \frac{c}{\lambda}$$

$$= (6.626 \cdot 10^{-34} \text{ J} \cdot \text{s}) \cdot \frac{\left(\dfrac{300 \text{ Mm}}{\text{s}} \right)}{450 \text{ nm}}$$

$$= 440 \cdot 10^{-21} \text{ J} = 440 \text{ zJ}$$

The seconds and meters canceled, leaving joule, the unit of energy. We substituted the prefix zepto (z) for 10^{-21}. To solve this problem in a calculator, we enter

6.626 $\boxed{\text{E}}$ $\boxed{+/-}$ 34 $\boxed{\times}$ 300 $\boxed{\text{E}}$ 6 $\boxed{\div}$
450 $\boxed{\text{E}}$ $\boxed{+/-}$ 9 $\boxed{=}$

Problem Set 15
Frequency

1. What is the reciprocal of time called?

2. Define frequency with an equation.

3. Define the hertz with an equation.

4. Give the wave law in an equation.

5. Give the wave law for radiation in vacuum.

6. What is the frequency range of audible sounds?

7. What kind of energy is produced when electricity accelerates?

8. What is the frequency (in hertz) of an ocean wave that has a period of 15 s?

9. What is the frequency (in hertz) of a phenomenon that has a period of 65 ns?

10. If the distance between wave crests (wavelength) is 12 m and the waves pass at a frequency of 350 mHz, what is the speed of the waves?

11. If the speed of sound is 330 m/s, what is the wavelength of a 250 Hz thunderclap?

12. If the speed of sound is 330 m/s, what is the frequency of a sound with a wavelength of 28 mm?

13. What is the frequency of 750 nm light (in vacuum)?

14. What is the wavelength of 55 EHz radiation (in vacuum)?

15. What is the frequency of 200 mm radio waves (in vacuum)?

16. What do you call radiation with a wavelength of 30 μm?

17. What color is 650 THz radiation?

18. What is the energy (in joules) of a photon of 5 EHz X rays?

19. What is the energy of a photon of orange light (wavelength 620 nm)?

CHAPTER 16

Angles

16.1 SUPPLEMENTARY UNITS

Recall from Section 2.3 that the two SI angle units are placed in a separate class called *supplementary units*, which the designers of SI regard as "dimensionless derived units." Angles are ratios, or fractions. They are actually pure numbers, not physical quantities like the others discussed thus far. Angular units are not essential in a coherent system like SI, but they are convenient in certain applications (such as angular velocity and radiant intensity).

16.2 THE RADIAN

The radian *(RAID-ian)* (rad) is the SI unit of plane, or two-dimensional, angle. It is defined by the ratio of the circumference of a circle to its radius, which is 2π. A radian is the angle cut off at the center of a circle by a radius' length (r) on the circumference. (See Figure 16.1.) A radian is therefore $1/2\pi$ of a circle. A complete circle has 2π rad (about 6.28 rad). Prefixes may be attached to form multiples such as milliradian (mrad) and microradian (μrad).

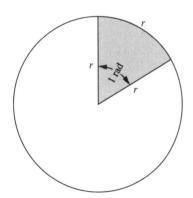

FIGURE 16.1 Radian.

16.3 THE STERADIAN

The steradian (sr) is the SI unit of solid, or three-dimensional, angle. A steradian is a square radian, or $1/4\pi$ of a sphere. In other words, it is the solid angle at the center of a sphere that cuts off an area on the surface of the sphere equal to a square radius (r^2). (See Figure 16.2.) A complete sphere therefore has 4π sr or about 12.6 sr. The steradian is used in the definition of the light units—the candela, lumen, and lux (Section 19.1).

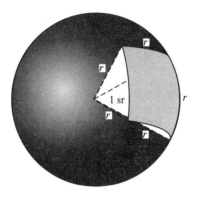

FIGURE 16.2 Steradian.

The advantage of radians and steradians is that they may cause the constant π (pi) to cancel out of equations. Because π is an irrational number it cannot be expressed exactly and is inconvenient in computation and measurement.

However, in surveying, mapmaking, navigation, and many other common applications where π is not involved, the radian and steradian are impractical because they introduce π into every measurement. In these cases, the more familiar unit, the degree, may be used with SI.

16.4 DEGREES AND RELATED UNITS (NON-SI)

Degrees (°) were invented thousands of years ago by Babylonian astrologer-priests. The Babylonians used a 360 day calendar, which led them to divide circles into 360 parts. (See Figure 16.3.) They knew a year actually had $365\frac{1}{4}$ days but 360 was apparently more convenient in their number system. However, degrees are very cumbersome in the decimal (base 10) number system we use today. Degrees are not SI, but because they are used worldwide they have been approved for use with SI when the radian is impractical.

Like hours of time, degrees are subdivided into 60 minutes (′) and the minutes are subdivided into 60 seconds (″). These ancient angle units are sometimes called *arc degrees, arc minutes,* and *arc seconds* to avoid confusing them with units of time or temperature. Scientific calculators have a key (usually labeled H.MS or D.MS) that converts decimal degrees to degrees, minutes, and seconds or vice versa.

Minutes and seconds of angle should be used only in cartography, navigation, astronomy, and related fields where they

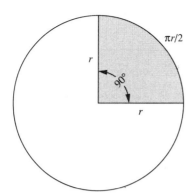

FIGURE 16.3 Degrees.

are hopelessly entrenched. In other cases you should use decimal degrees. A degree of latitude, measured north-south along a meridian, is about 111 km (a value obtained by dividing the 40 Mm circumference of Earth by 360°). A minute of latitude, or $\frac{1}{60}$ of 111 km, is called a *nautical mile* (nmi or NM) and defined as 1852 meters exactly. A nautical mile per hour is called a *knot* (kt). These units are acceptable for use in navigation although they are not SI.

16.5   GRADS (NON-SI)

In the original metric system, devised by the French in the eighteenth century, circles were divided into 400 parts called *grads* (or grades), with prefixed subdivisions (centigrad, etc.). A right angle is a convenient 100 grads. (See Figure 16.4.) A centigrad of latitude is 1 km. Grads were a great improvement over the cumbersome 360 degrees of the Babylonians. But when the metric system was developed, the British ruled the seas and were charting the world. They were not about to use any units invented by their arch rival, France. Now we have billions of maps with latitude and longitude measured in degrees, as well as compasses, surveying instruments, and countless other devices scaled in degrees. Grads are still used

by some Europeans but they are not SI and are *not* approved for use with SI.

Most scientific calculators offer the option of displaying angles in degrees, radians, or grads. They will also convert from one unit to another.

16.6   GRADIENTS (SLOPES)

Small angles from the horizontal such as the gradient of a highway, railroad, or river; the slope of a hillside; or the pitch of a roof are usually expressed as the vertical difference divided by the horizontal difference, or "rise per run." In SI, gradients are usually given in meters per kilometer (m/km). Because the meters cancel, this is really a pure number or "dimensionless unit" equivalent to 10^{-3}. Slopes are also commonly expressed in percent ($\% = 10^{-2}$).

We can convert between rise-per-run expressions and angle units (degrees or radians) by using trigonometry. In Figure 16.5, the tangent of angle θ equals the "rise," or elevation difference (h), over the "run," or horizontal distance (d).

$$\tan \theta = \frac{h}{d}$$

Therefore,

$$h = d \tan \theta \quad \text{and} \quad \theta = \text{atan}\left(\frac{h}{d}\right)$$

The function "atan" means arctangent, or "the angle whose tangent is." It is also called *inverse tangent* ($\tan^{-1}$). Scientific calculators have a key for it.

Example Problem 16.1.   Express a slope of 45 m/km in degrees.

Solution.   Use the last equation above.

$$\theta = \text{atan}\left(\frac{h}{d}\right) = \text{atan}\left(\frac{45}{1000}\right) = 2.6°$$

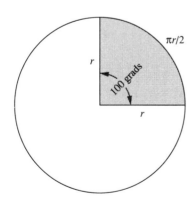

FIGURE 16.4 Grads.

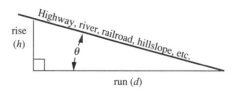

FIGURE 16.5 Gradient.

Other Non-SI Angle Units

right ascension (astronomy)	angular percent	meters per kilometer	quadrant
mil (military)	percent slope	pitch (roofs)	sextant
point (compasses)	feet per mile	octant	

Problem Set 16
Angles

1. Name the SI unit of plane angle.
2. What is the Babylonian unit of plane angle?
3. Name the original metric unit of plane angle (not SI).
4. What is the SI unit of solid angle?
5. How many radians are there in a circle?
6. How many steradians are there in a sphere?
7. Convert 233° to SI (radians).
8. Convert 2.84 rad to degrees.
9. Convert 75 grads to degrees.
10. Convert 35°16'40" to decimal degrees.
11. Convert 18.3478° to degrees, minutes, and seconds.
12. Convert 124°20'19" to SI.
13. Convert 15" (arc seconds) to a fraction of a circle.
14. What distance is 35' (arc minutes) of latitude, in nautical miles?
15. What distance is 3° of latitude, in kilometers?
16. Convert 12 knots to SI (m/s).
17. What fraction of a sphere is 8 sr?
18. Convert a stream gradient of 130 m/km to degrees.
19. Convert a 3% highway slope to radians.
20. Convert a roof pitch of 3 in 12 (3/12) to degrees.
21. Convert a gradient of 230 feet per mile to radians.

CHAPTER 17

Electromagnetic Units

17.1 CHARGE: THE COULOMB

Electromagnetic quantities have always been measured in metric units, so there are no inch-pound competitors for the SI units described in this chapter. Charge (Q) is the quantity of electricity. There are two kinds of electricity: positive charge (+), carried by protons, and negative charge (−), carried by electrons. We may regard the charge on a single proton or electron as the smallest possible amount of electricity, or "elementary charge" (e). But it is far too small to be a convenient unit for most purposes. Therefore, the SI unit of charge is the coulomb (C) (pronounced *KOO-lome* or *KOO-lahm*), which is approximately $6.241\,506 \cdot 10^{18}$ elementary charges.

$$C \approx 6.241\,506 \cdot 10^{18}\, e \quad \textit{(Elementary charges in a coulomb)}$$

This value was determined experimentally; it is not a definition. The coulomb had been the original and logical base unit for electromagnetic quantities. However, the ampere of current was adopted as the SI base unit (see next section). A coulomb is now defined as an ampere second.

$$C = A \cdot s \quad \textit{(Definition of coulomb)}$$

The coulomb was named after French physicist Charles de Coulomb (1736–1806), who discovered the relationship between charge and force, Coulomb's Law:

$$F_e = \frac{kQ_1Q_2}{r^2} \quad \textit{(Coulomb's Law)}$$

F_e is the electromagnetic force (in newtons) between two point charges in vacuum, Q_1 and Q_2 are the two charges (in coulombs), r is the distance between them (in meters), and k is a constant ($\approx 8.9876 \cdot 10^9$ N·m²/C²). The force F_e between unlike charges (+ and −) is attractive, and the force between like charges is repulsive. Note that Coulomb's Law is analogous to the Law of Gravitation (Section 10.2), with charge instead of mass.

Example Problem 17.1. What is the charge, in coulombs, of an electron or proton (the "elementary charge," e)?

Solution. Find the inverse of the number of elementary charges in a coulomb (above).

$$C = 6.241\,506 \cdot 10^{18}\, e$$

$$e = \frac{C}{6.241\,506 \cdot 10^{18}}$$
$$= 160.2177 \cdot 10^{-21}\ C$$
$$\approx 160\ zC$$

To simplify the answer, we substituted the prefix zepto (z) for 10^{-21}.

Example Problem 17.2. What is the electromagnetic force between the proton and the electron in a hydrogen atom if the distance between them is 53 pm?

Solution. Solve Coulomb's Law for force, using the value of the elementary charge determined in Example Problem 17.1. Remember to enter all prefixes.

$$F_e = \frac{kQ_1Q_2}{r^2}$$
$$= \frac{\left(\dfrac{8.99 \cdot 10^9\ N \cdot m^2}{C^2}\right) \cdot 160\ zC \cdot 160\ zC}{(53\ pm)^2}$$
$$= 82 \cdot 10^{-9}\ N$$
$$= 82\ nN$$

All the units canceled except newton, the unit of force. We simplified the answer by substituting the prefix nano (n) for 10^{-9}. We solve this problem in a calculator by entering

8.99 $\boxed{E}$ 9 $\boxed{\times}$ 160 $\boxed{E}$ $\boxed{+/-}$ 21 $\boxed{y^x}$ 2

$\boxed{\div}$ 53 $\boxed{E}$ $\boxed{+/-}$ 12 $\boxed{y^x}$ 2 $\boxed{=}$

Note that this force attracting the electron and proton in a hydrogen atom, 82 nanonewtons, is astonishingly large for such small particles. (Compare it to the gravitational force attracting the same two particles.)

17.2 ELECTRIC CURRENT: THE AMPERE

Electric current (I) is charge per time—the "flow rate" of electricity.

$$I = \frac{Q}{t} \quad \textit{(Definition of current)}$$

The SI unit of current is therefore a coulomb per second, which is called an ampere (A) (pronounced *AM-pir*).

$$A = \frac{C}{s} \quad \textit{(Original definition of ampere)}$$

The ampere was named after French physicist André Ampère [*ahm-PARE*] (1775–1836), who discovered the relationship between current and force. (Avoid the nickname "amp" for the ampere, especially in writing.) The ampere is the SI base unit for electromagnetic quantities. Its official modern definition is "that constant current which, if maintained in two straight parallel conductors of infinite length, of negligible circular cross section, and placed 1 meter apart in vacuum, would produce between those conductors a force of 200 nanonewtons per meter of length."

You can feel as little as 1 milliampere (mA) flowing through your body. A current of 100 mA is likely to kill you. An ordinary 100 W lightbulb draws slightly less than 1 A. Fuses and circuit breakers are electrical safety valves designed to shut off the current if it becomes too great. In a direct current (dc), the electricity flows in only one direction. In an alternating current (ac) it changes direction. Household current in the United States alternates direction 60 times a second, or 60 Hz (ac). Avoid the redundant expression "flow of current," since current *is* the flow of electricity.

17.3 ELECTRIC POTENTIAL: THE VOLT

Electric potential, commonly called voltage (*V*), is power per current or energy per charge. This quantity is also called *electromotive force* (emf) and symbolized *E*.

$$V = \frac{P}{I} = \frac{E}{Q} \quad \textit{(Definition of voltage)}$$

We obtained the second definition from the first by recalling that power is energy per time (*E/t*) and current is charge per time (*Q/t*). The time *t* cancels out, leaving energy per charge (*E/Q*).

The SI unit of electric potential is therefore a watt per ampere, or joule per coulomb. For convenience this derived unit is given the special name volt (V), after Italian physicist Alessandro Volta (1745–1827), who developed the battery.

$$V = \frac{W}{A} = \frac{J}{C} \quad \textit{(Definition of volt)}$$

The volt thus measures the power in a given flow of electricity or the potential energy stored in each bit of electricity. It helps to imagine an electric current through a wire as if it were water flowing through a hose. Current is analogous to the flow rate of the water. Voltage is analogous to the water pressure. High-voltage electricity will arc (spark or jump) farther through the air just as high-pressure water will squirt farther from a hose.

Electricity must have about 50 V to overcome the resistance of your dry skin. That explains why you do not feel a shock from a 12 V car battery even though it can deliver hundreds of amperes. Ordinary house current in the United States has a potential of 120 V. A typical bolt of lightning has more than 100 MV.

Extraneous information must not be attached to the symbol for volts (V) or to any other SI unit symbol. It must be clearly separated with a space. For example, an alternating current of 120 volts may be written 120 V (ac), but *not* 120 Vac or 120 VAC.

Example Problem 17.3. What power is produced when a 12 V car battery delivers a current of 150 A?

Solution. Solve the definition of voltage for power, substituting the definition W/A for volt.

$$V = \frac{P}{I}$$
$$P = VI$$
$$= 12 \text{ V} \cdot 150 \text{ A}$$
$$= 1.8 \cdot 10^3 \text{ V} \cdot \text{A}$$
$$= 1.8 \cdot 10^3 \left(\frac{W}{A}\right) \cdot \text{A}$$
$$= 1.8 \text{ kW}$$

We substituted the prefix kilo (k) for 10^3.

17.4 ELECTRIC RESISTANCE: THE OHM

Resistance *(R)* is voltage per current (*R = V/I*). This equation, called Ohm's Law, is usually written in the form

$$V = IR \quad \textit{(Ohm's Law)}$$

The SI unit of resistance is therefore a volt per ampere. For convenience this derived unit is given the special name ohm (Ω).

$$\Omega = \frac{V}{A} \quad \textit{(Definition of ohm)}$$

The ohm was named after German physicist Georg Ohm (1787–1854), who discovered the law. The symbol for ohm (Ω) is a Greek capital O (omega), since a roman O might be confused with zero.

Resistance can be imagined as a sort of "electrical friction." Water flowing through a long hose loses pressure because of the friction between the water and hose walls. Similarly, the current in a long wire loses voltage because of the resistance of the wire. An object's resistance, in ohms, tells us how many volts are needed to force an ampere of current through it. Resistance varies with the substance. Materials of low resistance, which let electricity flow easily, are called *conductors*. Materials of high resistance are called *insulators*. Metals are usually good conductors because they have lots of free, loose electrons. Electrical wires are made of conductors and are usually surrounded by insulators to keep the electricity from escaping.

Example Problem 17.4. If a 120 V lamp draws 800 mA of current, what is its resistance?

Solution. Solve Ohm's Law for resistance, and substitute the given values. Enter the prefix milli on milliampere, or move the decimal point three places left and enter it as 0.8 A.

$$V = IR$$

$$R = \frac{V}{I} = \frac{120 \text{ V}}{0.8 \text{ A}} = 150 \text{ V/A} = 150 \text{ }\Omega$$

17.5 OTHER ELECTROMAGNETIC UNITS

There are six other SI electromagnetic units, which will be described only briefly because they are limited to specialized fields. These and other specialized SI units are given in Table 4 on the inside back cover.

• The *siemens* (S) (pronounced *SEE-mens*) is the unit of electric conductance (G), or ease of flow—the reciprocal of resistance. A siemens is therefore a reciprocal ohm or ampere per volt (S = 1/Ω = A/V). It was formerly called a mho (*ohm* spelled backwards). The siemens was named after German-English inventor William Siemens (1823–1883).

• The *farad* (F) (pronounced *FAIR-ad*) is the SI unit of capacitance, or "electrical storage capacity" (C). It is a coulomb per volt (F = C/V), that is, a coulomb of electricity stored for each volt of potential between the plates of the capacitor (electrical storage device). The farad was named after English scientist Michael Faraday (1791–1867), who pioneered in the study of electromagnetism.

• The *weber* (Wb) is the SI unit of magnetic flux, or "amount of magnetism." A weber is a volt second (V·s)—the magnetic flux that produces a potential of 1 volt in a turn of wire as the flux is uniformly reduced to zero in 1 second. The weber, named after German physicist Wilhelm Weber (1804–1891), replaces the obsolete metric unit maxwell (Mx). A maxwell equals 10^{-8} Wb.

• The *tesla* (T) is the SI unit of magnetic flux density (B). It measures the intensity of a magnetic field (magnetic flux per area) in webers per square meter (T = Wb/m^2). The tesla, named after Croatian-American physicist and inventor Nikola Tesla (1856–1943), replaces the obsolete metric unit gauss (Gs). A gauss equals 10^{-4} T.

• The *henry* (H), named after American physicist Joseph Henry (1797–1878), is the unit of electromagnetic inductance (L). A changing magnetic field induces a voltage, and therefore a current, in a conductor. A henry is a weber of flux per ampere of current induced (H = Wb/A).

CHAPTER 18

The Mole

18.1 AMOUNT OF SUBSTANCE: THE MOLE

The mole (mol) is the newest SI base unit, adopted in 1971. In some ways it is an unnecessary complication of SI. A mole is essentially a pure number, approximately $6.022\,137 \cdot 10^{23}$, also known as Avogadro's Number. The mole is used for counting atoms, molecules, ions, formula units, electrons, or other elementary particles that make up substances. These particles are so small that we can only work with them in huge, bulk numbers. There is nothing special about the particular number $6.02 \cdot 10^{23}$, the mole. It is an accident of history.

18.2 ATOMIC MASS UNITS (NON-SI)

In the early nineteenth century, after the kilogram had been established as the standard of mass, matter was found to be made of atoms. Because elements always combine to form compounds in certain definite proportions, chemists were able to measure the *relative* masses of different kinds of atoms (elements). An oxygen atom is about 16 times more massive than a hydrogen atom, a carbon atom is about 12 times the mass of a hydrogen atom, and so on. However, for many years no one had any idea what the actual mass of an atom was in kilograms.

So a second mass unit was invented, called an *atomic mass unit*. The modern or unified atomic mass unit (symbol u) is arbitrarily defined as $\frac{1}{12}$ the mass of a carbon-12 atom (^{12}C). This is a convenient standard that gives the simplest atom, hydrogen, a mass of about 1 u. Because an ordinary hydrogen atom consists of only one proton and one electron and because electrons have very little mass, a proton also has a mass of about 1 u. Likewise, a neutron, which you can imagine as a proton and electron fused together, has a mass of about 1 u. So the *relative* mass of a substance (measured in atomic mass units), rounded to the nearest whole number, equals the number of nucleons (protons and neutrons) it contains. The relative atomic mass of each element (in u) is listed in the periodic table of the elements, found in many reference books and science texts.

It was not until quite recently that we learned, to any great precision, the value of an atomic mass unit in kilograms:

$$u \approx 1.660\,540 \cdot 10^{-27}\,\text{kg} \approx 1.66\,\text{yg} \quad \text{(yoctograms)}$$

Why do we need two different mass units? We do not. We could list atomic masses in yoctograms. We continue to use the non-SI atomic mass unit (u) partly out of tradition and partly because it tells us the number of nucleons.

Example Problem 18.1. A molecule of oxygen (O_2) has a relative mass of 32 u. What is its mass (in kilograms)?

Solution. Use the conversion factor above (1.66 yg/u).

$$32\,\text{u} \cdot \frac{1.66\,\text{yg}}{\text{u}} = 53\,\text{yg}$$

The atomic mass units canceled, leaving yoctograms (yg = 10^{-24} g = 10^{-27} kg).

18.3 DEFINING THE MOLE

Then what is a mole? It is simply the conversion factor between grams (g) and atomic mass units (u). It is the *number of atomic mass units in a gram*.

$$\text{mol} = \frac{\text{g}}{\text{u}} \approx \frac{\text{g}}{1.660\,540\,\text{yg}} \approx 6.022\,137 \cdot 10^{23}$$

(Definition of mole)

Since the unit (grams) cancels, the mole is really a pure number or "dimensionless" unit. Prefixes may be attached to form multiples such as millimole (mmol $\approx 6.02 \cdot 10^{20}$) and kilomole (kmol $\approx 6.02 \cdot 10^{26}$).

The official definition of the mole incorporates all of these ideas in an indirect way. A mole is formally defined as the amount of substance which has as many elementary particles as there are atoms in 12 grams of carbon-12.

18.4 USING MOLES

The mole is used to convert between atomic mass units (listed in the periodic table) and grams, which can be measured on a laboratory scale. Rearranging the equation in Section 18.3 we get

$$u = \frac{\text{g}}{\text{mol}}$$

Thus the relative mass of a substance (in u) equals the grams per mole of that substance. This fraction, g/mol, is used frequently in physics and chemistry to convert from grams to moles, and vice versa.

Why do we need these conversions? Is it not the purpose of SI to eliminate conversions? If we were developing SI from scratch today, we would probably base the unit of mass on an atom of some common element, instead of a cylinder of metal locked in a vault in France (the International Prototype Kilogram). We could do away with both atomic mass units and moles. Of course, we would still need to work with atoms and molecules in huge numbers. But instead of having a special name for that arbitrary number (the mole), we could use regular SI prefixes.

The mole can be used only with *pure substances*—elements, compounds, or elementary particles. You must specify both the substance and the kind of particle or group of particles (atom, ion, molecule, electron, formula unit, etc.). For example, you must say "2 mol of oxygen atoms" or "2 mol of oxygen molecules." You cannot measure something in moles unless you know its chemical composition (formula). You cannot measure mixtures of substances, such as rocks, wood, or living things, in moles. So the mole is not a convenient unit for everyday use. But it is essential in science.

Example Problem 18.2. The relative masses of hydrogen (H) and oxygen (O) atoms are 1.01 u and 16.0 u, respectively. What is the mass (in kilograms) of 2 mol of water molecules (H_2O)?

Solution. The relative mass of a water molecule is the sum of the relative masses of its atoms.

$$(2 \cdot 1.01 \text{ u}) + 16.0 \text{ u} = 18.0 \text{ u}$$

Multiplying this relative mass by 2 moles and substituting g/mol for u, we get

$$2 \text{ mol} \cdot \frac{18 \text{ g}}{\text{mol}} = 36 \text{ g}$$

The moles canceled, leaving grams.

Problem Set 18
The Mole

For Problems 1–6, round the numbers to be given to three significant digits.

1. State the number of atoms in 1 mol of nitrogen atoms.

2. State the number of molecules in 1 mol of nitrogen molecules.

3. State the number of electrons in 1 mol of electrons.

4. State the number of atoms in 1 mmol of hydrogen atoms.

5. State the number of formula units in 1 µmol of sodium chloride.

6. What is the number of ions in 1 kmol of calcium ions?

7. Define the atomic mass unit (u).

8. What is the value of an atomic mass unit (u) in SI units (kilograms)?

9. What is the relative mass of a proton (in u), rounded to one significant digit?

10. What is the relative mass of a neutron (in u), rounded to one significant digit?

11. What is the relative mass of a hydrogen atom (^{1}H), in u (rounded to one significant digit)?

12. Define the mole with an equation.

Use the following relative atomic masses to solve Problems 13–19.

hydrogen (H)	1.01 u	oxygen (O)	16.00 u
carbon (C)	12.00 u	sodium (Na)	22.99 u
chlorine (Cl)	35.45 u	uranium (U)	238 u

Give the mass (in kilograms) of the following.

13. A carbon atom (^{12}C)

14. 1 mol of oxygen atoms

15. 1 mol of salt (NaCl)

16. 1 mol of water (H_2O)

17. 3 mol of water

18. 6 kmol of uranium atoms

19. How many moles of hydrogen atoms are in 5 g of hydrogen atoms?

20. How many atoms are there in 2 mol of carbon atoms?

CHAPTER 19

Physiological Units

There are six specialized SI units that measure the impacts of certain kinds of radiant energy (radiation) on humans. They are summarized in Table 4, on the back inside cover. In many respects these units are unnecessary complications of SI. Most are not truly coherent—that is, they are not defined as simple products or quotients of the most fundamental base units.

19.1 LIGHT

Light is radiant energy (electromagnetic radiation) with wavelengths visible to the human eye, about 400 to 700 nanometers. Light can be measured with the same units as any other kind of energy: joules, watts, watts per square meter, and so on. However, in some fields light is also measured in the units described in this section—the candela, lumen, and lux. These units were established in the nineteenth century, before it was possible to measure light in absolute terms with respect to other kinds of energy. Now, incorporated into SI, the light units are defined by radiation at a frequency of 540 terahertz, a greenish color (wavelength about 555 nm) to which human eyes are most sensitive. For other frequencies (or wavelengths) one must multiply by experimentally determined "weighting factors" that take the eye's varying sensitivity into account. So unlike most SI units, the light units do not have simple, absolute definitions.

All three of the light units measure quantities popularly called "brightness." However, there is a subtle difference in the way they are measured.

• The *candela* (cd) (pronounced [*can-DELL-uh*]) is the SI base unit of luminous intensity (I). It was originally defined in the nineteenth century as the brightness of a specified type of candle. The modern definition is $\frac{1}{683}$ of a watt per steradian (W/sr) of monochromatic light measured at a frequency of 540 THz." Thus the candela measures the brightness of a light source in *a given direction* (solid angle). In the United States the candela, which means "candle" in Latin, is popularly called a candle or candlepower, but these are not correct SI terms.

• The *lumen* (lm) is the SI unit of luminous flux. It equals $\frac{1}{683}$ of a watt measured at 540 THz but is officially defined as a candela steradian.

$$\text{lm} = \text{cd} \cdot \text{sr} = \frac{1}{683} \text{ W at 540 THz} \quad \textit{(Definition of lumen)}$$

The lumen measures the total light flow, or light power, in *all directions*. Lightbulbs are rated in lumens. A conventional 100 W incandescent bulb emits about 1700 lm. Most of the 100 W of electrical energy flowing into the bulb flows out as heat (infrared radiation), not light, so it would be more accurate to call it a "heatbulb."

For historical reasons, the candela is the base unit for light. But it is more logical to think of a candela as a lumen per steradian (cd = lm/sr). Since there are 4π steradians in a sphere (Section 16.3), a 1 cd source that is emitting light uniformly in all directions gives off 4π, lumens, or about 12.6 lm.

• The *lux* (lx) is the SI unit of illuminance, or illumination (E). It is a lumen per square meter, or $1/683$ W/m² measured at 540 THz.

$$\text{lx} = \text{lm/m}^2 \quad \textit{(Definition of lux)}$$

Building codes specify a minimum illumination in lux. The light sensitivity of video cameras is also rated in lux. Illumination of about 1000 lx (or 1 klx) is generally desirable for reading a book. The word is unchanged in the plural (say "10 lux," not "10 luxes").

Example Problem 19.1. The beam of an 8000 cd spotlight shines through a solid angle of 3 sr. How many lumens does the spotlight emit?

Solution. By definition, lm = cd · sr. Therefore

$$8000 \text{ cd} \cdot 3 \text{ sr} = 24\,000 \text{ cd} \cdot \text{sr} = 24\,000 \text{ lm} = 24 \text{ klm}$$

Example Problem 19.2. In an otherwise dark room, a 150 lm flashlight beam is focused on 3 m² of wall. What is the illumination of the wall (in lux).

Solution. By defintion, lx = lm/m². Therefore

$$\frac{150 \text{ lm}}{3 \text{ m}^2} = 50 \text{ lm/m}^2 = 50 \text{ lx}$$

19.2 IONIZING RADIATION

Ionizing radiation is very energetic radiation with the ability to ionize atoms and possibly damage living cells. The derived units used to measure it have been given special names so that it will not be mistaken for more benign forms of radiation.

• Radioactivity is the frequency of nuclear decay, that is, the number of nuclear transformations or disintegrations per time ($1/t$). It is therefore measured in decays per second ($1/s$). To distinguish radioactivity from other frequencies measured in hertz, the SI unit of radioactivity was given the special name *becquerel* (Bq) (pronounced *BECK-er-uhl*).

$$\text{Bq} = \frac{1}{\text{s}} \qquad \textit{(Definition of becquerel)}$$

Radiation is emitted during nuclear decay, so radioactivity can be measured with a Geiger counter or similar device. Radioactive elements (radioisotopes or radionuclides) have nuclei that spontaneously decay at a constant rate, depending on the substance. The becquerel is therefore an indirect measure of the mass of radioactive material present, if the substance is known. It was named for French physicist Antoine-Henri Becquerel (1852–1908), who discovered radioactivity in 1896. The 1986 explosion at the Chernobyl nuclear power plant in Ukraine is estimated to have released about 2 EBq (exabecquerels) into the environment. The becquerel replaces the noncoherent unit curie (1 curie = 37 GBq).

Example Problem 19.3. The radioactivity of natural uranium is 13 TBq/kg. A sample of material known to contain natural uranium has a radioactivity of 6 MBq. What mass of uranium is present?

Solution. If we divide Bq by Bq/kg, the becquerels will cancel leaving kilogram (the unit of mass). We must enter the prefixes mega and tera.

$$6 \text{ MBq} \cdot \left(\frac{\text{kg}}{13 \text{ TBq}} \right) = \frac{6 \cdot 10^6 \text{ Bq} \cdot \text{kg}}{13 \cdot 10^{12} \text{ Bq}}$$

$$= 460 \cdot 10^{-9} \text{ kg} = 460 \text{ μg}$$

• The *gray* (Gy) is the SI unit of absorbed dose (D). It is defined as a joule of ionizing radiation absorbed per kilogram of body mass (Gy = J/kg). The gray, named after English radiobiologist Louis Harold Gray (1905–1965), replaces the obsolete metric unit, rad (1 rad = 0.01 Gy = 10 mGy).

• The *sievert* (Sv) is an index of potential damage to human health. It measures the "effective dose," or "dose equivalent" (H), that is, the absorbed dose weighted for the type of radiation involved. It indicates how much of the energy taken into the body (J/kg) is actually absorbed by the cells. The sievert is defined as the absorbed dose in grays, multiplied by a "quality factor" Q, which varies with the radioactive substance, times any other factor N which may be stipulated by the International Commission on Radiological Protection.

$$\text{Sv} = \text{Gy} \cdot Q \cdot N \qquad \textit{(Definition of sievert)}$$

The sievert was named after Swedish radiobiologist Rolf M. Sievert (1896–1966). Because it includes statistically determined weighting factors, the sievert is not coherent; it is not defined as a simple product or quotient of the base units. In fact, it is the only SI unit with no absolute definition at all. It was adopted, despite objections, because of the critical importance of radiation to human health and the possibility that effective dose might be confused with the total absorbed dose in grays. The average American is exposed to about 2 mSv (millisieverts) per year of natural, background radiation. A typical chest X ray is about 0.1 mSv. A dose of several sieverts in one day is almost certain to kill you. The sievert replaces the obsolete metric unit, rem (1 rem = 0.01 Sv = 10 mSv).

Problem Set 19
Physiological Units

1. The SI light units are defined by monochromatic radiation measured at what frequency?

2. What is the SI unit of light power (the flow rate of visible radiation)?

3. What is the SI unit that measures the light power flowing through a given solid angle?

4. What are two popular names for the candela, widely used in the United States but not correct in SI?

5. If the light beam from a flashlight has an angle of 2 sr and a luminous intensity of 500 cd, what is the total light power (luminous flux) emitted by the flashlight (in lumens)?

6. If a 1700 lm lightbulb emits light uniformly in all directions, what is the luminous intensity of the bulb (in candelas)?

7. If a 2 klm beam of light is spread uniformly over 8 m² of wall, what is the mean illumination on the wall (in lux)?

8. If a video camera requires an illumination of at least 10 lx to operate, what minimum light power (luminous flux), in lumens, will be needed to illuminate a 5 m² surface?

9. If radium spontaneously decays at a rate of 37 TBq/kg, what mass of radium is present when the radioactivity is 18 MBq?

10. If a 65 kg woman absorbs 1.5 J of ionizing radiation, what is her total absorbed dose (whole-body mass), in grays?

Convert the following to SI.

11. Radioactivity of 20 picocuries per liter

12. A dose equivalent of 40 millirems

13. An absorbed dose of 300 millirads

CHAPTER 20

Compound Units

Compound units are derived units without special names. An unlimited number of them can be formed by combining other units. Common examples are listed below, alphabetically by quantity.

A quantity may have several different names. The word *specific* in a quantity usually means "per mass." *Molar* usually means "per mole." *Rate* usually means "per time." *Density* usually means "per volume" or "per area." Although a quantity has only one SI unit, the same unit may measure several different quantities. For example, the unit joule per kelvin (J/K) measures both heat capacity (C_p) and entropy (S).

Quantity	Unit symbol	Unit name
Absorbed dose rate	Gy/s	gray per second
Acceleration, angular (α)	rad/s^2	radian per second squared
Acceleration, linear (a)	m/s^2	meter per second squared
Bending moment or torque (τ)	N·m	newton meter
Coefficient of heat transfer	W/(m^2·K)	watt per square meter kelvin
Concentration, mass	kg/m^3	kilogram per cubic meter
Concentration, molar (molarity) (M)	mol/L	mole (of solute) per liter (of solution)
Concentration (molality) (m)	mol/kg	mole (of solute) per kilogram (of solvent)
Conductivity (electric) (σ)	S/m	siemens per meter
Current density (J)	A/m^2	ampere per square meter
Density (mass) (D, ρ)	kg/m^3	kilogram per cubic meter
Discharge (streamflow) (Q)	m^3/s	cubic meter per second
Electric charge density (ρ)	C/m^3	coulomb per cubic meter
Electric dipole moment (p)	C·m	coulomb meter

Quantity	Unit symbol	Unit name
Electric field strength (E)	V/m	volt per meter
Electric flux density (D)	C/m^2	coulomb per square meter
Energy density	J/m^3	joule per cubic meter
Energy, specific	J/kg	joule per kilogram
Entropy (S)	J/K	joule per kelvin
Entropy, specific or internal	J/(kg·K)	joule per kilogram kelvin
Entropy, molar	J/mol·K	joule per mole kelvin
Exposure (to X or gamma rays)	C/kg	coulomb per kilogram (of body mass)
Flow (mass per time)	kg/s	kilogram per second
Flow (volume per time)	m^3/s	cubic meter per second
Force per length	N/m	newton per meter
Fuel consumption, specific	kg/J	kilogram per joule
HEAT (Thermal Quantities)		
Heat capacity (C_p)	J/K	joule per kelvin
Heat capacity, molar	J/(mol·K)	joule per mole kelvin
Heat capacity, specific (c_p)	J/(kg·K)	joule per kilogram kelvin
Heat density	J/m^3	joule per cubic meter
Heat flux density (q)	W/m^2	watt per square meter
Latent heat (ΔQ)	J/kg	joule per kilogram
Thermal conductance (C)	W/(m^2·K)	watt per square meter kelvin
Thermal conductivity (k, λ)	W/(m·K)	watt per meter kelvin
Thermal diffusivity (a)	m^2/s	square meter per second

Quantity	Unit symbol	Unit name
Thermal insulance	$K \cdot m^2/W$	kelvin square meter per watt
Thermal resistance	K/W	kelvin per watt
Thermal resistivity	$K \cdot m/W$	kelvin meter per watt
Impulse	$N \cdot s$	newton second
Insolation	J/m^2	joule per square meter
Insulation (thermal) (R)	$K \cdot m^2/W$	kelvin square meter per watt
Intensity (I)	W/m^2	watt per square meter
Irradiance (E)	W/m^2	watt per square meter
Irradiation	J/m^2	joule per square meter
Luminance (L)	cd/m^2	candela per square meter
Luminous energy	$lm \cdot s$	lumen second
Magnetic field strength (H)	A/m	ampere per meter
Mass per area (fabric, paper, etc.)	kg/m^2	kilogram per square meter
Mass per length (wire, rope, etc.)	kg/m	kilogram per meter
Molality (m)	mol/kg	mol per kilogram (of solvent)
Molar energy	J/mol	joule per mole
Molarity (M)	mol/L	mole per liter (of solution)
Moment of force (M)	$N \cdot m$	newton meter
Moment of inertia (static) (I)	$kg \cdot m^2$	kilogram meter squared
Moment of inertia (with rotation)	$kg \cdot m^2/rad^2$	kilogram meter squared per radian squared
Moment of momentum (static)	$kg \cdot m^2/s$	kilogram meter squared per second
Moment of momentum (with rotation)	$kg \cdot m^2/(rad \cdot s)$	kilogram meter squared per radian second

Quantity	Unit symbol	Unit name
Momentum, linear (p)	$kg \cdot m/s$	kilogram meter per second
Momentum, angular (L)	$kg \cdot m^2/s$	kilogram meter squared per second
Permeability (magnetic) (μ)	H/m	henry per meter
Permittivity (ϵ)	F/m	farad per meter
Power density	W/m^2	watt per square meter
Radiance (L)	$W/(m^2 \cdot sr)$	watt per square meter steradian
Radiant intensity (N)	W/sr	watt per steradian
Reluctance (R)	$1/H$	per henry
Resistivity (ρ)	$V \cdot m$	ohm meter
Specific heat (c_p)	$J/(kg \cdot K)$	joule per kilogram kelvin
Specific energy	J/kg	joule per kilogram
Specific entropy	$J/(kg \cdot K)$	joule per kilogram kelvin
Specific volume	m^3/kg	cubic meter per kilogram
Speed (v)	m/s	meter per second
Surface tension (γ)	N/m	newton per meter
Thermal (see HEAT)		
Thrust per mass	N/kg	newton per kilogram
Torque (static) (τ)	$N \cdot m$	newton meter
Torque (with rotation)	$N \cdot m/rad$	newton meter per radian
Transmissivity	m^2/s	square meter per second
Velocity, angular (ω)	rad/s	radian per second
Velocity, linear (v)	m/s	meter per second
Viscosity, dynamic or absolute (η)	$Pa \cdot s$	pascal second
Viscosity, kinematic (ν)	m^2/s	square meter per second
Wave number (σ)	$1/m$	per meter

Rules for the Correct Use of SI

1. CAPITALIZATION OF UNITS

Units are common nouns and are therefore not capitalized. Do not capitalize a spelled-out unit or prefix (even if its symbol is capitalized). The word *Celsius* is the only exception.

Correct	Incorrect
100 watts (100 W)	100 Watts
20 newtons (20 N)	20 Newtons
35 megajoules (35 MJ)	35 Megajoules
20 degrees Celsius	20 degrees celsius

2. CAPITALIZATION OF SYMBOLS

Units named after a person and the larger prefixes (mega and up) have capitalized symbols. So does liter (L), to avoid confusion with the numeral 1. All other SI symbols are always lowercase, regardless of the surrounding text.

Correct	Incorrect	
km (kilometer)	KM	Km
mm (millimeter)	MM	Mm
mL (milliliter)	ML	ml
mg (milligram)	Mg	MG
Mg (megagram)	MG	

3. PUNCTUATION

Never use a period after a symbol (unless it ends a sentence).

Correct	Incorrect
25 kg	25 kg.
15 cm	15 cm.
60 s	60 sec.

4. PLURALS

Use normal English plurals when spelling out or pronouncing unit names. The hertz, lux, and siemens are unchanged in the plural. Do not add an *s* to a symbol to indicate plural.

Correct	Incorrect
25 kilograms	25 kgs
10 centimeters	10 cms
5 millihenries	5 millihenrys
0.3 meter	0.3 meters

Correct	Incorrect
60 hertz	60 hertzes
50 lux	50 luxes
20 siemens	20 siemenses

5. SPACING

Always leave a space between the number and the unit for legibility. On computers use a "hard space," which does not expand when text is full-justified or break when a word wraps to the next line.

Correct	Incorrect
30 L	30L
21 m	21m
95 lm (lumens)	95lm
20 °C	20°C

6. LETTER STYLE

Use upright (normal, or roman) letters for SI symbols, regardless of the surrounding text. Use slanted (*italic, or oblique*) letters for quantity symbols. Correct examples:

3 g   (3 grams)

3 *g*   (3 times the acceleration of Earth's gravity)

7. SPELLING

The Germanic spellings, *mete*r and *liter*, are the most common in the United States and have been adopted by the U.S. government. However, the original and internationally preferred spellings, used in all English-speaking countries except the United States, are *metre* and *litre*. The Spanish spellings are *metro* and *litro*. The spelling of other units and prefixes is nearly the same in all nations that use the Roman alphabet. SI *symbols*, on the other hand, are the same in all languages and alphabets.

8. PRONUNCIATION

The primary stress is always on the first syllable (whether it is a prefix or plain unit) so that the sound and identity of the prefix will not be lost. The only exceptions are the candela (*can-DELL-uh*) and steradian (*ste-RAID-ian*).

	Correct	Incorrect
kilometer (km)	*KILL-oh-meter*	*kil-LOM-muh-ter*
pascal (Pa)	*PASS-kuhl*	*pass-KAHL*
kilopascal (kPa)	*KILL-oh-pass-kuhl*	*kill-o-pass-KAHL*

9. NEVER ABBREVIATE

Use the correct symbols, never abbreviations.

Correct		Incorrect
s	(second)	sec
m^2	(square meter)	sq. meter
cm^3	(cubic centimeter)	cc
km/h	(kilometer per hour)	kph
A	(ampere)	Amp

10. NO MIXING OF SYMBOLS AND NAMES

Do not mix symbols with spelled-out names in the same expression. Do not use symbols alone in text (without a number).

Correct	Incorrect
MHz (megahertz)	MHertz
5 kg/L	5 kg/liter
5 kilograms per liter	5 kilograms/liter
joules per kilogram	J per kg
about a meter	about a m
about 1 m	

11. NO ATTACHMENTS

Do not attach extraneous information to a unit symbol. Spell it out or abbreviate it, separated from the unit by a space.

Correct	Incorrect
120 V, alternating current	120 VAC
120 V (ac)	120 V(ac)
120 V ac	120 Vac
100 kPa (gage)	100 kPag
50 m^3 (STP)	50 std m^3

12. LEADING ZEROES

Put a zero in front of a leading decimal point so the point will not be overlooked.

Correct	Incorrect
0.3 m	.3 m

13. POWERS OF TEN

Avoid mixing prefixes with powers of ten in the same expression. That is confusing and redundant since the prefix *is* a power of ten.

Correct	Poor
10^5 m	10^7 cm

Exception: Use powers of ten with kilogram, not with gram (because kilogram is the base unit).

Correct	Poor
10^2 kg	10^5 g

14. PREFIX CHOICES

So that quantities will be easy to read and indicate the approximate precision, try to use a prefix that gives a number between 0.1 and 1000.

Good	Poor
8 km	8000 m
50 mL	0.05 L

Exceptions: When quantities are arranged in a column in a table they should all have the same prefix, even if the range is large. For areas or volumes it may be necessary to use several placeholding zeroes.

15. POWERS OF UNITS

When raising units to a power, use exponents in the normal way. For areas or volumes only, the word *square* or *cubic* precedes the unit name. In all other cases, the word *squared* or *cubed* follows the name in the usual fashion. Correct examples:

$$hm^3 = (hm)^3 = \text{cubic hectometer } (volume)$$
$$cm^2 = (cm)^2 = \text{square centimeter } (area)$$
$$m/s^2 = \text{meter per second squared } (acceleration)$$

16. PREFIXES HECTO, DEKA, DECI, AND CENTI

To avoid needless complication of SI and to simplify calculation, the four prefixes that are not multiples of 1000 should be avoided except in areas, volumes, and the centimeter of length (see Chapters 5, 6, and 7 for details).

17. DECIMAL POINTS AND DIGIT SEPARATORS

In SI you may use either a period or a comma for the decimal marker. Many countries use a comma, which is easier to see and write hastily and cannot be mistaken for the multiplication dot ($\cdot$). In those countries, periods are traditionally used as digit separators. Since this is the opposite of the English-language convention, great confusion can result. Therefore, the preferred international practice, and that recommended by the U.S. government media style guide (*American National Standard for Metric Practice;* see Bibliography), is to separate groups of digits with spaces rather than punctuation. Spaces are inserted every three digits, counting both left and right from the decimal point. On computers use a "hard space," which does not expand when full-justifying or wrap to the next line. In printed material, the space should be about the width of the letter *i* . *Exception:* the space is usually omitted if there are *only* four digits to the left or right of the decimal point.

Recommended SI	U.S. (English)
1203.56	1,203.56
1 203.56	
1203,56	
1 203,56	
5 000 000	5,000,000
0.0047	.0047
0.004 7	
0,0047	
0,004 7	
3.768 923	3.768923
3,768 923	

18. MULTIPLICATION

To multiply symbols use a raised dot (·) but do not pronounce it or spell it out.

Correct	Incorrect	
15 newton meters	15 newton-meters	
15 N·m	15 N-m	15 N × m
	15 Nm	15 N m

19. DIVISION

To divide symbols, use a horizontal bar, slash, or negative exponents. Use the word *per* when pronouncing or spelling out divided units.

Correct	Incorrect
J/kg	joules/kg
$\dfrac{J}{kg}$	J ÷ kg
J·kg^{-1}	$\dfrac{joules}{kg}$
joules per kilogram	joules over kilogram

20. COMPLEX DIVISION

Do not use more than one slash or "per" in an expression unless parentheses or commas are inserted to eliminate ambiguity.

Correct equivalent expressions

joules per kilogram kelvin

joules per kilogram, per kelvin

J/(kg·K)

(J/kg)/K

$\dfrac{J}{kg·K}$

J·kg^{-1}·K^{-1}

Incorrect expressions

joules per kilogram per kelvin

J/kg/K

The two incorrect expressions are ambiguous. They could mean either (J/kg)/K, as in the correct example, or J/(kg/K), which is not the same.

21. COMPOUND UNITS

In a compound unit, use a prefix only with the left-hand or numerator unit. Otherwise it will be difficult to compare quantities.

Good	Poor
kN·m	N·mm
g/m^2	g/cm^2

Exceptions: The kilogram is used, even though it has a prefix, because it is the base unit. Density is often expressed in mass per liter or mass per cubic centimeter (milliliter). See Chapter 8.

Good	Poor
kJ/kg	J/g

22. NEVER USE FRACTIONS OR MIX PREFIXES

Express numbers with decimals, not fractions. Do not use more than one prefix with a unit.

Correct	Incorrect
6.5 kg	6½ kg
2.65 m	2 m 65 cm
12 nm	12 mμm

APPENDIX **B**

Conversion Factors for Non-SI Units

B.1 SIMPLE CONVERSIONS

There are no conversions in pure SI because it has only one unit for each quantity. However, you will often need to convert non-SI data into SI or vice versa. This appendix defines five hundred non-SI units alphabetically by name and/or symbol.

- To convert a quantity to SI, *multiply* by the number in the SI definition column (the conversion factor).
- To convert from SI to non-SI, *divide* by the number in the SI definition column.

Be sure to round your answers to the same precision as your original data (see Chapter 3 for rounding rules).

Example Problem B.1. Convert 28 miles to SI.

Solution. We look up "mile" in the alphabetical list. We select "mile (international)," which the boldface type indicates is the most common of the three "miles" used in the United States. It is defined as exactly 1.609 344 kilometers (* indicates an exact definition). We do not need such great precision so we round it to 1.6 and multiply by 28 to get 45 km.

B.2 FACTOR LABEL METHOD

For more complex conversions it is best to set up the conversion on paper, using the "factor label method." A unit definition can be expressed as an equation or as a conversion fraction that equals 1. For example, the alphabetical list tells us that a foot (international) is defined as exactly 0.3048 meter.

$$ft = 0.3048 \text{ m}$$

We can rearrange this into a conversion fraction by dividing both sides by foot (ft).

$$\frac{0.3048 \text{ m}}{ft} = 1$$

Since a conversion fraction always equals 1, we may invert it without altering it:

$$\frac{ft}{0.3048 \text{ m}} = 1$$

Either fraction above may be pronounced "0.3048 meter per foot." The numerical factor always goes with the SI unit, because non-SI units are defined in terms of SI (not vice versa).

To convert a quantity into different units, multiply or divide by an appropriate conversion fraction(s) so that all the units cancel except those you want. To divide by a fraction, invert it and multiply. The *quantity* remains unchanged because multiplying or dividing by 1 does not alter an expression. Only the *units* are altered.

Example Problem B.2. Convert 550 light years to SI.

Solution. We look up "light year" and express its definition as a fraction.

$$\frac{9.46 \text{ Pm}}{ly}$$

We multiply the given distance, 550 ly, by this fraction, 9.46 petameters per light year, entering the prefix peta (P = 10^{15}) on the calculator.

$$550 \text{ ly} \cdot \frac{9.46 \cdot 10^{15} \text{ m}}{ly} = 5.2 \cdot 10^{18} \text{ m} = 5.2 \text{ Em}$$

The light years (ly) canceled, leaving meters. To simplify the answer we substituted the prefix exa for 10^{18}.

Example Problem B.3. Convert 5.8 acre feet per day to SI.

Solution. We look up the definitions of acre foot and day (mean solar), express them as fractions, and arrange them with our original data so that all the non-SI units cancel. To make them cancel, we must invert (divide by) the fraction 86 400 s/d. For neatness, we will set up the problem in fraction bar notation (Section 2.6).

$$\frac{5.8 \text{ af}}{d} \left| \frac{1233 \text{ m}^3}{af} \right| \frac{d}{86\,400 \text{ s}} = 83 \cdot 10^{-3} \text{ m}^3/\text{s} = 83 \text{ L/s}$$

We substituted the alternative name liter (L) for 10^{-3} m³.

Example Problem B.3. Convert 1250 Btu (International table) to calories (thermochemical).

Solution. In this case we are converting from one non-SI unit to another. The list does not usually give conversion factors between two non-SI units. That would be impractical since there are thousands of possible combinations. However, we can use SI as an intermediate step or "bridge." We look up the definitions of Btu (International table) and calorie (thermochemical). Then we arrange them as conversion fractions, along with our original data, so that all the units cancel except calorie. We must invert the fraction 4.184 J/cal.

$$\frac{1250 \text{ Btu}}{} \cdot \frac{1055 \text{ J}}{\text{Btu}} \cdot \frac{\text{cal}}{4.184 \text{ J}} = 315 \cdot 10^3 \text{ cal}$$

A few units, such as degrees Fahrenheit of temperature, are defined by an equation rather than a simple factor. These equations are given in the SI definition column. To convert, follow the equation.

B.3 NON-SI SYSTEMS

Most non-SI units evolved by custom and do not belong to any coherent system. However, there are many partial systems, some of which are listed in Table B.1 and noted in parentheses after the unit names in the alphabetical list.

Units in different systems often have the same name, even when their values are very different. For example, Imperial (British) gallons, quarts, pints, and tons are much larger than the U.S. versions. Table B.1 may help you identify the unit you want. For example, it tells you that precious metals (such as gold, silver, and platinum) are measured in the troy ounce, which is slightly larger than the common avoirdupois ounce.

Some units with the same name measure completely different quantities. There is a pound of mass (454 g), a "pound" of force (4.4 N), a "pound" of pressure (6.9 kPa), and a "pound" of paper (3.75 g/m^2). Be sure to select the correct unit for the quantity you are measuring.

TABLE B.1 Some Non-SI Unit Systems	
System	**Comments**
British absolute	An engineering system based on the poundal, foot, and second.
Imperial (English)	Formerly used in all English-speaking countries except the United States (Britain, Canada, Australia, Ireland, New Zealand, etc.). Now largely replaced by SI.
Inch-pound (U.S.)	General term for non-SI units traditionally used in the United States. Most are based on the 1959 yard and pound (mass).
U.S. dry	Volume of fruits, grains, and vegetables.
U.S. liquid	Volume of most commodities.
U.S. survey	Length and area units based on the 1893 yard. Used for precise geodetic land surveying.
International (English)	Length, mass, and area units defined in 1959 to reconcile the slightly different U.S. and Imperial versions. Do not confuse with the International System (SI).
Mass	
Avoirdupois	Most things.
Troy or Apothecary	Precious metals and medicines.
Old Versions of the Metric System	
CGS	Base units are centimeter, gram, and second.
emu	Electromagnetic CGS units (prefixed ab-).
esu	Electrostatic CGS units (prefixed stat-).
MKpS (MKfS)	Base units are meter, kilopond (kilogram-force), and second.
MTS	Base units are meter, ton, and second.
Time	
Mean solar	Everyday units based on Earth's rotation relative to the sun.
Sidereal	Astronomical units based on Earth's rotation relative to the stars.

Alphabetical List of Non-SI Units and Symbols

Nonalphabetic and Greek characters are listed at the end, following the letter *Y*.

KEY

*	Exact SI definitions (others are rounded to a precision appropriate for the unit).
≈	The conversion factor is determined experimentally or theoretically and is therefore not exact (not a definition).
boldface	The most common units in ordinary U.S. use.
EIA	U.S. Energy Information Administration.
EU	European Union.
USDA	U.S. Department of Agriculture.

Non-SI unit [Symbol]	Quantity	SI definition
abampere (CGS-emu)	current	10* A
abcoulomb (CGS-emu)	charge	10* C
abfarad (CGS-emu)	capacitance	1* GF
abhenry (CGS-emu)	inductance	1* nH
abmho (CGS-emu)	conductance	1* GS
abohm (CGS-emu)	resistance	1* nΩ
abvolt (CGS-emu)	potential	10* nV
acre (commercial) (36 000 square feet) [Ac]	area	3344.5 m^2
acre (U.S. survey) (43 560 square U.S. survey feet) [Ac]	area	4046.873 m^2
acre foot [af]	volume	1233.5 m^3
acre foot per year [af/y]	flow	39.09 · 10^{-6} m^3/s
acre inch [ac in.]	volume	102.8 m^3
agate (printing) (1/14 in.)	length	1.814 286 mm
"amp" (slang for ampere)		
ampere hour [A·h]	charge	3600* C
angstrom [Å]	length	0.1* nm
angular percent (1/100 circle) [A%]	plane angle	π/50* rad
annum (year) (365 $\frac{1}{4}$ d) [a]	time	31.56 Ms
arc degree (1/360 circle) [°]	plane angle	π/180* rad
arc minute (1/21 600 circle) [']	plane angle	π/10 800* rad
arc second (1/1 296 000 circle) ["]	plane angle	π/648 000* rad
are (obsolete metric) [a]	area	100* m^2
astronomical unit [AU]	length	≈149.5979 Gm
atmosphere (standard) [atm]	pressure	101.325* kPa
atmosphere (technical) (kgf/cm^2) [at]	pressure	98.0665* kPa
atomic mass unit (unified, carbon-12) [u]	mass	≈1.660 54 · 10^{-27} kg ≈1.660 54 yg
bale (cotton) (500 pounds)	mass	227 kg
balthazar (champagne)	volume	12* L
bar [b]	pressure	100* kPa
barleycorn (1/3 inch)	length	8.467 mm
barn [b]	area	100* fm^2
barrel (Imperial) (36 Imp. gallons) [bbl]	volume	0.163 659 m^3
barrel (petroleum) (42 U.S. gallons) [bbl]	volume	0.158 987 m^3
barrel (U.S. dry) (105 dry quarts) [bbl]	volume	0.115 628 m^3
barrel (U.S. federal) (31 gallons) [bbl]	volume	0.117 348 m^3
barrel (U.S. federal proof spirits) (40 U.S. gallons) [bbl]	volume	0.151 416 m^3
barrel (U.S. liquid) (31.5 U.S. gallons) [bbl]	volume	0.119 240 m^3
barrel of crude petroleum (EIA mean)	mass	≈136 kg
barrel of LNG (liquid natural gas) (EIA mean)	energy (gross)	≈3.17 GJ
barrel of petroleum (consumption, EIA 1990)	energy (gross)	≈5.77 GJ
barrel of petroleum (production, EIA 1990)	energy (gross)	≈6.15 GJ
barye (CGS) (dyn/cm^2)	pressure	0.1* Pa
Bcf (billion cubic feet)	volume	28.3169 · 10^6 m^3
biot	current	10 A
bit (computers) (short for "*bi*nary di*git*")		
board foot (1/12 ft^3) [bf]	volume	2.3597 · 10^{-3} m^3
bottle (standard wine)	volume	0.750* L
British thermal unit (EU) (International Table) [Btu$_{IT}$]	energy	1055.056 J
British thermal unit (mean) [Btu$_m$]	energy	1055.87 J
British thermal unit (thermochemical) [Btu$_{th}$]	energy	1054.350 J
British thermal unit (39 °F) [Btu$_{39°}$]	energy	1059.67 J
British thermal unit (59 °F) [Btu$_{59°}$]	energy	1054.804 J

Non-SI unit [Symbol]	Quantity	SI definition
British thermal unit (60 °F) [Btu$_{60°}$]	energy	1054.68 J
Btu (*see* British thermal unit)		
Btu$_{IT}$·ft/(h·ft^2·°F)	thermal conductivity	1.730 735 W/(m·K)
BTUH (*see* Btu per hour)		
Btu$_{IT}$·in/(h·ft^2·°F)	thermal conductivity	0.144 227 9 W/(m·K)
Btu$_{IT}$ in/(s ft^2·°F)	thermal conductivity	519.2204 W/(m·K)
Btu$_{IT}$ per cubic foot [Btu$_{IT}$/ft^3]	energy density	37.258 95 kJ/m^3
Btu$_{IT}$ per degree Fahrenheit [Btu$_{IT}$/°F]	heat capacity	1899.101 J/K
Btu$_{IT}$ per degree Rankine [Btu$_{IT}$/°R]	heat capacity	1899.101 J/K
Btu$_{IT}$ per hour [Btu$_{IT}$/h]	power	0.293 071 1 W
Btu$_{IT}$ per hour [Btu$_{th}$/h]	power	0.292 875 1 W
Btu$_{IT}$/(h·ft^2·°F)	thermal conductance	5.678 263 W/(m^2·K)
Btu$_{th}$ per minute [Btu$_{th}$/min]	power	17.572 W
Btu$_{IT}$ per pound [Btu$_{IT}$/lb]	specific energy	2.326* kJ/kg
Btu$_{IT}$/(lb·°F)	specific heat	4186.8* J/(kg·K)
Btu$_{IT}$ per second [Btu$_{IT}$/s]	power	1055.056 W
Btu$_{IT}$/(s·ft^2·°F)	thermal conductance	20.441 75 kW/(m^2·K)
Btu$_{IT}$ per square foot [Btu$_{IT}$/ft^2]	heat density	11.356 53 kJ/m^2
Btu$_{IT}$/(ft^2·h)	heat flow density	3.154 591 W/m^2
Btu$_{IT}$/(ft^2·s)	heat flow kW/m^2 density	11.356 53
bushel (US dry) (2150.42 in^3) [bu]	volume	0.035 239 m^3 35.239 L
bushel (Imperial) (8 Imperial gallons) [bu]	volume	36.369 L
bushel of barley (USDA) [bu]	mass	21.8 kg
bushel of corn (shelled) (USDA) [bu]	mass	25.4 kg
bushel of oats (USDA) [bu]	mass	14.5 kg
bushel of wheat, potatoes, or soybeans (USDA) [bu]	mass	27.2 kg
butt (U.S. liquid)	volume	0.476 96 m^3
byte (B) (unit of computer memory corresponding to one		

Non-SI unit [Symbol]	Quantity	SI definition
C (roman numeral for hundred)	(pure number)	100*
c (speed of radiation in vacuum)	velocity	299 792 458* m/s
cable (720 ft) [ca]	length	219.456* m
caliber (0.01 inch) [cal]	length	0.254* mm
calorie (international table) [cal$_{IT}$]	energy	4.1868* J
calorie (large or kilogram) (*see* kilocalorie)		
calorie (mean) [cal$_m$]	energy	4.190 02 J
calorie (nutrition) (Cal) [kcal$_{th}$]	energy	4184* J
calorie (thermochemical) [cal$_{th}$]	energy	4.184* J
calorie (15 °C or normal) [cal$_{15°}$]	energy	4.185 80 J
calorie (20 °C) [cal$_{20°}$]	energy	4.181 90 J
calorie (1950 CIPM) [cal]	energy	4.1855* J
cal$_{th}$/(cm·s·°C)	thermal conductivity	418.4* W/(m·K)
calorie per gram [cal$_{th}$/g]	specific energy	4184* J/kg
cal$_{th}$/(g·°C)	specific heat	4184* J/(kg·K)
cal$_{th}$/(g·K)	specific heat	4184* J/(kg·K)
calorie per minute [cal$_{th}$/min]	power	69.733 33 mW
calorie per second [cal$_{th}$/s]	power	4.184* W
calorie per square centimeter [cal$_{th}$/cm^2]	irradiation	41.84* kJ/m^2
cal$_{th}$/(cm^2·min)	irradiance	697.3333 W/m^2
cal$_{th}$/(cm^2·s)	irradiance	41.84* kW/m^2
candela per square inch [cd/in^2]	luminance	1550.003 cd/m^2
candle (c) (obsolete name for candela [cd])		
candlepower (cp) (obsolete name for candela [cd])		
carat (metric) [ct]	mass	0.2* g
cc (obsolete abbreviation for cubic centimeter [cm^3], or milliliter [mL])		
CCF (100 cubic feet)	volume	2.831 685 m^3
centigrade degree (obsolete name for degree Celsius [°C])		
centiliter [cL]	volume	10* mL
centimeter of mercury [cm Hg]	pressure	1.333 22 kPa
centimeter of water (ISO) [cm H$_2$O]	pressure	98.0665* Pa
centipoise [cP]	viscosity	0.001* Pa·s
centistokes [cS]	kinematic viscosity	1* · 10^{-6} m^2/s
cf (*see* cubic foot)		

Non-SI unit [Symbol]	Quantity	SI definition
cfm (*see* cubic foot per minute)		
cfs (*see* cubic foot per second)		
chain (66 U.S. survey feet) [ch]	length	20.116 840 m
circular inch ($\pi\cdot(0.5$ inch$)^2$) [cir in.]	area	5.067 075 cm^2
circular mil ($\pi\cdot(0.5$ mil$)^2$) [cir mil]	area	506.7075 µm^2
clo	insulation	0.200 371 K·m^2/W
cord (mean wood, 20% water, 50% burning efficiency)	energy (net)	≈10^{-16} GJ
cord (solid wood, 5/8 nominal)	volume (net)	≈2.3 m^3
cord (nominal, stacked roundwood) (128 ft^3) [cd]	volume (gross)	3.6 m^3
cord-foot (nominal, 16 ft^3) [cd-ft]	volume (gross)	0.45 m^3
cubic foot (cu ft) [ft^3]	volume	0.028 316 85 m^3 28.316 85 L
cubic foot of natural gas (dry @ 60 °F and 14.73 psi) (EIA 1990)	energy (gross)	≈1.093 MJ
cubic foot per minute (cfm) [ft^3/min]	flow	471.95·10^{-6} m^3/s
cubic foot per second (cfs) [ft^3/s]	flow	0.028 316 9 m^3/s
cubic inch (cu in) [in^3]	volume	16.387 064* cm^3
cubic inch per minute [in^3/min]	flow	273.1177 mm^3/s
cubic meter of natural gas (mean @ STP)	energy	≈38.6 MJ
cubic meter of natural gas (mean @ STP)	mass	≈1.3 kg
cubic meter of crude petroleum (mean)	energy (gross)	≈36 GJ
cubic meter of crude petroleum (mean)	mass	≈860 kg
cubic meter of gasoline (EIA)	energy	≈35 GJ
cubic meter of hardwood (mean, 20% water @ 50% burning efficiency)	energy (net)	≈7 GJ
cubic meter of softwood (mean, 20% water @ 50% burning efficiency)	energy (net)	≈4.4 GJ
cubic mile (international) [mi^3]	volume	4.168 182 km^3
cubic yard (cu yd) [yd^3]	volume	0.764 555 m^3
cubic yard per minute [yd^3/min]	flow	0.012 742 6 m^3/s

Non-SI unit [Symbol]	Quantity	SI definition
cubit (English) (18 inches)	length	0.457 m
cup (1/16 U.S. gallon) [C]	volume	237 mL
curie [Ci]	radioactivity	37* GBq
"cusec" (slang for cubic foot per second)		
cwt (*see* hundredweight)		
cy (*see* cubic yard)		
day (mean solar) (24 h) [d]	time	86 400 s
day (sidereal) [d]	time	86 164.09 s
debye (CGS-esu)	electric dipole moment	3.3356·10^{-30} C·m
decibel (sound level) (*L*) (unbiased or C-range) [dB] where *p* is in pascals and *L* is in decibels	pressure amplitude	$p = 2\cdot 10^{(0.1L-5)}$
decibel (sound level) (*L*) (unbiased or C-range) [dB] where *I* is in W/m^2 and *L* is in decibels	intensity	$I = 10^{(0.1L-12)}$
deciliter [dL]	volume	100* mL
degree (1/360 circle) [°]	plane angle	π/180 rad
degree Celsius (interval) [°C]	temperature	1* K
degree Celsius (temperature) [°C]	temperature	$T_K = T_{°C} + 273.15$
degree centigrade (obsolete name for degree Celsius)		
degree Fahrenheit (interval) [°F]	temperature	5/9* K
degree Fahrenheit (temperature) [°F]	temperature	$T_K = (T_{°F} + 459.67)/1.8$ $T_{°C} = (T_{°F} - 32)/1.8$
°F·h/Btu$_{IT}$	thermal resistance	146.366 mK/W
°F·s/Btu$_{IT}$	thermal resistance	0.526 918 K/W
°F·h·ft^2/Btu$_{IT}$	thermal insulance	0.176 110 K·m^2/W
°F·h·ft^2/(Btu$_{IT}$·in)	thermal resistivity	6.933 471 K·m/W
degree Rankine [°R]	temperature	$T_K = T_{°R}/1.8$
denier (thread)	mass per length	0.111 111 mg/m
dram (apothecary) (60 grains) [dr ap]	mass	3.887 934 6* g
dram (avoirdupois) ($27\frac{11}{32}$ grains) [dr av]	mass	1.771 845 g
dram (Imperial fluid) (1/1280 Imp. gallon) [fl dr]	volume	3.551 634 mL
dram (U.S. fluid) (1/1024 U.S. gallon) [fl dr]	volume	3.696 691 mL
drum (55 U.S. gallons)	volume	0.208 m^3
dwt (pennyweight) (troy) (24 grains) [dwt]	mass	1.555 174 g

Non-SI unit [Symbol]	Quantity	SI definition
DWT (deadweight ton) (*see* ton, displacement)		
dyne (CGS metric) ($g \cdot cm/s^2$) [dyn]	force	10* μN
dyne centimeter [dyn·cm]	torque	100* μN·m
dyne per square centimeter [dyn/cm^2]	pressure	0.1* Pa
Einstein unit (radiant energy capable of photochemically changing 1 mole of a photosensitive substance)		
electromagnetic units (CGS-emu) (*see* ab-)		
electron volt [eV]	energy	$\approx$160.219 zJ
electrostatic units (CGS-esu) (*see* stat-)		
erg (CGS) (dyn·cm) [erg]	energy	100* nJ
erg per second [erg/s]	power	100* nW
erg/($cm^2 \cdot$s)	irradiance	1* mW/m^2
Eötvös unit (10^{-9} gal/cm)	horizontal gradient of acceleration due to gravity	1* $(nm/s^2)/m$
°F (*see* degree Fahrenheit)		
faraday (carbon-12) (gram-equivalent charge) [Fdy]	charge	96.4853 kC
fathom (6 U.S. survey feet) [fm]	length	1.8288 m
fbm (feet board measure) (*see* board foot)		
fermi (obsolete name for femtometer [fm])		
fifth (1/5 U.S. gallon)	volume	0.757 08 L
firkin (9 U.S. gallons)	volume	34.07 L
flask (mercury) (76 pounds)	mass	34.4730 kg
fluid units (*see* ounce [fluid] and dram [fluid])		
foot (British survey) (′) [ft]	length	0.304 799 7 m
foot (international) (′) [ft]	length	0.3048* m
foot (U.S. survey) (′) (1200/3937 meter) [ft]	length	0.304 800 61 m
foot of water (39.2 °F) [ft H_2O]	pressure	2988.98 Pa
foot per hour (fph) [ft/h]	speed	84.666 67 mm/s
foot per mile (gradient) [ft/mi]	plane angle	atan (*gradient*/5280)
foot per minute (fpm) [ft/min]	speed	5.08* mm/s
foot per second (fps) [ft/s]	speed	0.3048* m/s
foot per second squared [ft/s^2]	acceleration	0.3048* m/s^2
"foot-pound" (torque) (*see* pound-force foot)		
foot pound-force [ft · lbf]	energy	1.355 818 J
foot pound-force per hour [ft·lbf/h]	power	376.6161 μW

Non-SI unit [Symbol]	Quantity	SI definition
foot pound-force per minute [ft·lbf/min]	power	22.596 97 μW
foot pound-force per second [ft·lbf/s]	power	1.355 818 W
foot poundal [ft·pdl]	energy	42.140 11 mJ
foot ton-force [ft·tonf]	energy	2711.636 J
footcandle (lm/ft^2) [fc]	luminance	10.763 91 lx
footlambert [ftL]	luminance	3.426 259 cd/m^2
fph (*see* foot per hour)		
fpm (*see* foot per minute)		
fps (*see* foot per second)		
franklin	current	333.5641 pC
ft (*see* foot)		
ft^2 (*see* square foot)		
ft^3 (*see* cubic foot)		
ft^4	second moment of area	$8.630\,975 \cdot 10^{-3}\ m^4$
furlong (1/8 U.S. survey mile) [fur]	length	201.168 m
gage (Many different systems of length units called *gage* [or gauge] are used in industry. Within each system, the size of the unit varies slightly so there is no simple conversion factor. Consult reference tables.)		
g (standard free-fall) (acceleration due to gravity on Earth's surface) [*g*]	acceleration	9.806 65* m/s^2
gal (CGS) (cm/s^2) [Gal]	acceleration	10* mm/s^2
gallon (Imperial) [gal]	volume	4.546 092 L
gallon (U.S. dry) (1/8 bushel) [gal]	volume	4.404 884 L
gallon (U.S. liquid) (231 in^3) [gal]	volume	$3.785\,412 \cdot 10^{-3}\ m^3$ 3.785 412 L
gallon of butane (EIA std)	energy (gross)	$\approx$109 MJ
gallon of fuel oil (EIA std)	energy (gross)	$\approx$147 MJ
gallon of gasoline (EIA std)	energy (gross)	$\approx$133 MJ
gallon of propane (EIA std)	energy (gross)	$\approx$97 MJ
gallon (U.S.) per day (gpd) [gal/d]	flow	$43.812\,64 \cdot 10^{-9}\ m^3/s$
gallon per horsepower hour gal/(hp·h)	specific fuel consumption	$1.410\,089 \cdot 10^{-9}\ m^3/J$
gallon per minute (gpm) [gal/min]	flow	$63.090\,20 \cdot 10^{-6}\ m^3/s$
gamma (*γ*) (obsolete name for microgram [10^{-9} kg])		
gamma (*γ*) (obsolete name for nanotesla [nT])		
gauss (CGS) (Mx/cm^2) [Gs]	magnetic flux density	100* μT

Non-SI unit [Symbol]	Quantity	SI definition
gf (*see* gram-force)		
gigannum (10^9 years) [Ga]	time	31.56 Ps
gigawatt hour [GW·h]	energy	3.6* TJ
gilbert (CGS-emu) [Gi]	current	0.795 775 A
gill (Imperial) [gi]	volume	0.142 065 3 L
gill (U.S. liquid) (1/32 gal) [gi]	volume	0.118 294 L
gm (obsolete symbol for gram [g])		
gpd (*see* gallon per day)		
gpm (*see* gallon per minute)		
grad (grade) (1/400 circle)	plane angle	π/200 rad
grain (avoirdupois, troy, or apothecary) [gr]	mass	64.798 91* mg
grain per gallon (U.S.) [gr/gal]	density	17.118 06 mg/L
gram-force (MKpS) [gf]	force	9.806 65* mN
gram-force per square centimeter [gf/cm²]	pressure	98.0665* Pa
gram-mole (g-mol) (obsolete name for mole [mol])		
great gross (12 gross)	(pure number)	1728*
gross (12 dozen) [gr]	(pure number)	144*
gross register ton (GRT) (*see* ton, register)		
hand (horses) (4 inches)	length	10.2 cm
hartree	energy	4.360 aJ
hectoliter [hL]	volume	100* L
hogshead (U.S.) (63 gal) [hhd]	volume	0.2385 m³
horsepower (boiler) [hp]	power	9809.5 W
horsepower (electric) [hp]	power	746* W
horsepower (Imperial) [hp]	power	745.7 W
horsepower (mechanical) (550 ft·lbf/s) [hp]	power	745.6999 W
horsepower (MKpS metric) (*cheval vapeur*) (75 kgf·m/s) [cv]	power	735.5 W
horsepower (water) [hp]	power	746.043 W
hour (mean solar) (60 min) [h]	time	3600* s
hour (right ascension) (1/24 circle) [ʰ]	plane angle	261.7994 mrad
hour (sidereal) (1/24 sidereal day) [h]	time	≈3590.170 s
hundredweight, long (112 lb) [cwt]	mass	50.802 35 kg
hundredweight, short (100 lb) [cwt]	mass	45.359 24 kg
hyl (metric slug)	mass	9.806 65* kg

Non-SI unit [Symbol]	Quantity	SI definition
in² (*see* square inch)		
in³ (*see* cubic inch)		
in⁴	second moment of area	$416.2314 \cdot 10^{-9}$ m⁴
inch (″) (1/12 international foot) [in]	length	25.4* mm
inch, miner's (*see* miner's inch)		
inch of mercury (32 °F) [in. Hg]	pressure	3386.38 Pa
inch of mercury (60 °F) [in. Hg]	pressure	3377.85 Pa
inch of water (39.2 °F) [″WC]	pressure	249.082 Pa
inch of water (60 °F) [″WC]	pressure	248.84 Pa
inch per second (ips) [in/s]	speed	25.4* mm/s
inch per second squared [in/s²]	acceleration	25.4* mm/s²
in. Hg (*see* inch of mercury)		
ips (*see* inch per second)		
jeroboam (champagne)	volume	3* L
jigger (1.5 fl oz) [ji]	volume	45 mL
"k" (sports slang for kilometer [km])		
karat (ratio of gold in an alloy) [K]	(pure number)	1/24*
kayser	wave number	100* m⁻¹
kcal (*see* kilocalorie)		
keg ($\frac{1}{4}$ barrel, U.S. federal)	volume	29.3 L
kgf (*see* kilogram-force)		
"kilo" (slang for kilogram [kg])		
kiloannum [ka]	time	31.56 Gs
kilobyte [kB]	computer memory	2^{10}* bytes
kilocalorie (international table) [kcal$_{IT}$]	energy	4186.8* J
kilocalorie (mean) [kcal$_m$]	energy	4190.02 J
kilocalorie (1950 CIPM) [kcal]	energy	4185.5* J
kilocalorie (thermochemical) (=nutritional calorie) [kcal$_{th}$]	energy	4184* J
kilogram-calorie (*see* kilocalorie)		
kilogram-force (kilopond) (MKpS metric) (*g*·kg) [kgf]	force	9.806 65* N
kilogram-force meter [kgf·m]	torque	9.806 65* N·m
kilogram-force per square centimeter [kgf/cm²]	pressure	98.0665* kPa

Non-SI unit [Symbol]	Quantity	SI definition
kilogram-force per square meter [kgf/m^2]	pressure	9.806 65* Pa
kilogram-force per square millimeter [kgf/mm^2]	pressure	9.806 65* MPa
kilogram of butane	energy (gross)	≈49.4 MJ
kilogram of carbohydrate	energy (gross)	≈17 MJ
kilogram of coal equivalent (UN std) (7 · 10^6 cal$_{IT}$)	energy (gross)	29.3076* MJ
kilogram of coal (consumption, EIA 1990)	energy (gross)	≈24.9 MJ
kilogram of coal (production, EIA 1990)	energy (gross)	≈25.5 MJ
kilogram of crude petroleum (UN std, 1986)	energy (gross)	≈41.88 MJ
kilogram of fat (mean)	energy (gross)	≈38 MJ
kilogram of LNG (liquid natural gas) (UN weighted average)	energy (gross)	≈45 MJ
kilogram of natural gas (mean)	energy (gross)	≈52 MJ
kilogram of propane	energy (gross)	≈50 MJ
kilogram of wood (mean, 20% water)	energy (gross)	≈15 MJ
kiloliter [kL]	volume	1* m^3
kilometer per hour (kph) [km/h]	speed	0.277 778 m/s
kilopond [kp] (see kilogram-force)		
kiloton (10^6 kg) [kt]	mass	1* Gg
kiloton (nuclear weapons) [kt]	energy	4.184* TJ
kilowatt hour [kW·h]	energy	3.6* MJ
kilowatt hour, electrical (gross energy input of fossil fuels, hydropower, wind, photovoltaic, wood waste, or solar thermal consumed per net kW·h of electrical output; EIA 1990)		
[k·Wh$_e$]	energy	≈10.95 MJ
kilowatt hour, geothermal (gross geothermal energy input per net kW·h of electrical output; EIA 1990) [kWh$_g$]		
	energy	≈22.35 MJ
kilowatt hour, nuclear (gross nuclear energy input per net kW·h of electrical output; EIA 1990) [kWh$_n$]		
	energy	≈11.36 MJ
kilowatt hour, thermal (net electrical output) [kWh$_t$]	energy	3.6* MJ
kip (1000 lbf) [kip]	force	4448.222 N
kip per square inch [ksi]	pressure	6.894 757 MPa
"klik" (military slang for kilometer [km])		
knot (nautical mile per hour) [kt]	speed	0.514 444 m/s
kph (kilometer per hour) [km/h]	speed	0.277 778 m/s
ksi (see kip per square inch)		

Non-SI unit [Symbol]	Quantity	SI definition
lambda (λ) (obsolete name for microliter [μL], or cubic millimeter [mm^3])		
lambert ([10^4/π] cd/m^2) [L]	luminance	3183.099 cd/m^2
langley (cal$_{th}$/cm^2) [ly]	irradiation	41.84* kJ/m^2
lb (see pound, avoirdupois)		
lbf (see pound-force)		
lbm (see pound, avoirdupois)		
lbt (see pound, troy)		
league (U.S. statute) (3 mi)	length	4.828 km
league, marine (3 nmi)	length	5.556* km
light year (c·year) [ly]	length	9.461 Pm
link (U.S. survey) (0.01 ch) [li]	length	0.201 168 m
liter atmosphere (45° lat.) [L·atm]	energy	101.3 J
liter of gasoline (EIA)	energy	≈35 MJ
lumen per square foot [lm/ft^2]		10.7639 lx
M (roman numeral for thousand)	(pure number)	1000*
M (see molarity)		
m (see molality)		
mach number (speed of sound in air) (varies with temperature and pressure)	speed	≈330 m/s
Maf (million acre feet)	volume	1.233 489 km^3
magnum (champagne)	volume	1.5* L
maxwell (CGS) [Mx] (cm^2·Gs)	magnetic flux	10* nWb
MBF (1000 board feet)	volume	2.360 m^3
MCF (1000 cubic feet)	volume	28.316 85 m^3
mcg (incorrect symbol for microgram [μg])		
megabyte [MB]	computer memory	2^{20}* bytes
megannum (million years) [Ma]	time	31.6 Ts
megaliter [ML]	volume	1000* m^3
megaton (10^9 kg) [Mt]	mass	1* Tg
megaton (nuclear weapons) [Mt]	energy	4.184* PJ
megawatt hour [MW·h]	energy	3.6* GJ
methuselah (champagne)	volume	6* L
metric ton (tonne) [t]	mass	1000* kg
metric ton of coal equivalent (UN std) (7 · 10^9 cal$_{IT}$)	energy	29.3076* GJ
metric ton of crude petroleum (UN std, 1986)	energy	≈41.88 GJ

Non-SI unit [Symbol]	Quantity	SI definition
metric ton of uranium equivalent (mean net electrical output per metric ton of uranium, using conventional ^{235}U technology).	energy	≈140 TJ
mgd (million gallons per day)	flow	$43.813 \cdot 10^{-3}$ m³/s
mho (obsolete name for siemens [S], the reciprocal ohm)		
mi² (see square mile)		
microinch [μin.]	length	25.4* nm
micron (μ) (obsolete name for micrometer [μm])		
mil (1/6400 circle) [mil]	plane angle	981.7477 μrad
mil (thou) (0.001 inch) [mil]	length	25.4* μm
mile (international) (5280 ft) [mi]	length	1.609 344* km
mile (nautical) [nmi]	length	1.852* km
mile (U.S. survey) (5280 U.S. survey feet) [mi]	length	1.609 347 km
mile per gallon (U.S.) (mpg) [mi/gal]	fuel consumption	$235/mpg$ = L/100 km 0.425 km/L
mile per hour (mph) [mi/h]	speed	0.447 04* m/s 1.609 344* km/h
mile per minute [mi/min]	speed	26.8224* m/s
mile per second (mps) [mi/s]	speed	1609.344* m/s
millennium (1000 years)	time	31.6 Gs
millibar [mb]	pressure	100* Pa
millimeter of mercury (0 °C) (torr) [mm Hg]	pressure	133.322 Pa
millimicron (mμ) (obsolete name for nanometer [nm])		
miner's inch (1.2 ft³/min)	flow	$0.5663 \cdot 10^{-3}$ m³/s
miner's inch (1.5 ft³/min)	flow	$0.7079 \cdot 10^{-3}$ m³/s
minim (U.S. liquid) [min] (1/480 fluid ounce)	volume	61.61 μL
minute (arc minute) (1/60°) [']	plane angle	290.8882 μrad
minute (mean solar) [min]	time	60 s
minute (right ascension) (1/1440 circle) [ᵐ]	plane angle	4.363 323 mrad
minute (sidereal) [m]	time	≈59.836 17 s
MM (roman numeral for million)	(pure number)	$1* \cdot 10^6$
MMBF (million board feet)	volume	2360 m³
mm Hg (see millimeter of mercury)		
molality (moles of solute per kilogram of solvent) [m]	concentration	1* mol/kg
molarity (moles of solute per liter of solution) [M]	concentration	1* mol/L
month (lunar, phase, or synodic) (≈29.5306 d)	time	≈2.551 44 Ms
month (mean calendar) (30½ d) [mo]	time	2.63 Ms

Non-SI unit [Symbol]	Quantity	SI definition
month (sidereal) (≈27.3217 d)	time	≈2.3609 Ms
mpg (see miles per gallon)		
mph (see miles per hour)		
nautical mile [nmi]	length	1852* m
nebuchadnezzar (champagne)	volume	15* L
oersted (CGS) [Oe]	magnetic intensity	79.577 47 A/m
ohm centimeter	resistivity	10* mΩ·m
ohm circular mil per foot	resistivity	1.662 426 nΩ·m
ounce (avoirdupois) (437.5 grains) [oz]	mass	28.349 523 g
"ounce" (fabric) (see ounce per square yard)		
ounce, fluid (Imperial) (1/160 Imp. gallon) [fl oz]	volume	28.413 06 mL
ounce, fluid (U.S.) (1/128 U.S. gallon) [fl oz]	volume	29.573 53 mL
ounce (troy or apothecary) [ozt]	mass	31.103 48 g
ounce per cubic inch [oz/in³]	density	1.729 994 Mg/m³
ounce per gallon (Imperial) [oz/gal]	density	6.236 023 kg/m³
ounce per gallon (U.S.) [oz/gal]	density	7.489 152 kg/m³
ounce per square foot [oz/ft²]	mass per area	0.305 152 kg/m²
ounce per square yard [oz/yd²]	mass per area	33.905 75 g/m²
ounce-force (g·oz) [ozf]	force	0.278 013 85 N
ounce-force inch [ozf·in]	torque	7.061 552 mN·m
oz (see ounce, avoirdupois)		
ozf (see ounce-force)		
ozt (see ounce, troy)		
pace, geometrical (5 feet)	length	1.5 m
palm (3 inches)	length	7.62 cm
parsec (parallax arc second) [pc]	length	30.856 78 Pm
peck (U.S. dry) (1/4 bushel) [pk]	volume	8.809 768 L
pennyweight (troy) (24 grains) [dwt]	mass	1.555 174 g
percent slope (gradient) [%]	plane angle	atan (gradient/100)
percent, angular [A%]	plane angle	π/50 rad
perch or pole (see rod)		
perm (0 °C)		57.2135 ng/(Pa·s·m²)
perm (23 °C)		57.4525 ng/(Pa·s·m²)

Non-SI unit [Symbol]	Quantity	SI definition
perm inch (0 °C)		1.453 22 ng/(Pa·s·m)
perm inch (23 °C)		1.459 29 ng/(Pa·s·m)
pH (acid concentration) where $[H_3O^+]$ is the concentration of hydronium ions in mol/L	pH	$pH \approx -\log [H_3O^+]$
phot (CGS) (cd·sr/cm^2) [ph]	illuminance	10* klx
pica (1/6 French inch) [pca]	length	4.217 517 6* mm
pint (Imperial) (1/8 Imperial gallon) [pt]	volume	0.568 261 5 L
pint (U.S. dry) (1/64 bushel) [pt]	volume	0.550 610 5 L
pint (U.S. liquid) (1/8 U.S. gallon) [pt]	volume	0.473 176 5 L
pitch (roof) (/12)	plane angle	atan (*pitch*/12)
point (jewelers) (0.01 carat) [pt]	mass	2* mg
point (printing) (1/12 pica) [pt]	length	0.351 459 8* mm
poise (CGS) (dyn·s/cm^2) [P]	viscosity	0.1* Pa·s
pole (*see* rod)		
pound (avoirdupois) (#) (poundmass) (7000 grains) [lb]	mass	0.453 592 37* kg
pound (troy or apothecary) (5760 grains) [lbt]	mass	0.373 241 7 kg
pound of force (*see* pound-force)		
"pound" of paper (*see* sub)		
"pound" of pressure (*see* pound-force per square inch)		
pound foot squared [lb·ft^2]	moment of inertia	42.140 11 g·m^2
pound inch squared [lb·in^2]	moment of inertia	292.6397 mg·m^2
pound per cubic foot [lb/ft^3]	density	16.018 46 kg/m^3
pound per cubic yard [lb/yd^3]	density	0.593 276 4 kg/m^3
pound per cubic inch [lb/in^3]	density	27.6799 Mg/m^3
pound per foot hour [lb/ft·h]	viscosity	413.3789 µPa·s
pound per foot second [lb/ft·s]	viscosity	1.488 164 Pa·s
pound per gallon (Imperial) [lb/gal]	density	99.776 37 kg/m^3
pound per gallon (U.S.) [lb/gal]	density	119.8264 kg/m^3
pound per horsepower hour [lb/(hp·h)]	specific fuel consumption	168.9659 µg/J
pound per hour [lb/h]	mass per time	125.9979 mg/s
pound per minute [lb/min]	mass per time	7.559 873 g/s
pound per second [lb/s]	mass per time	0.453 592 4 kg/s
pound per square foot [lb/ft^2]	mass per area	4.882 428 kg/m^2

Non-SI unit [Symbol]	Quantity	SI definition
"pound per square inch" (psi) (*see* pound-force per square inch)		
poundal (British absolute) (lb·ft/s^2) [pdl]	force	0.138 254 95 N
pound-force (*g*·lb) [lbf]	force	4.448 222 N
pound-force foot [lbf·ft]	torque	1.355 818 N·m
pound-force foot per inch [lbf·ft/in]		53.378 66 N·m/m
pound-force inch [lbf·in]	torque	0.112 985 N·m
pound-force inch per inch [lbf·in/in]		4.448 222 N·m/m
pound-force per foot [lbf/ft]	force per length	14.5939 N/m
pound-force per inch [lbf/in]	force per length	175.1268 N/m
pound-force per pound [lbf/lb]	thrust per mass	9.806 65* N/kg
pound-force per square foot (psf) [lbf/ft^2]	pressure	47.880 26 Pa
pound-force per square inch (psi) [lbf/in^2]	pressure	6.894 757 kPa
pound-force second per square foot [lbf·s/ft^2]	viscosity	47.880 26 Pa·s
pound-force second per square inch [lbf·s/in^2]	viscosity	6.894 757 kPa·s
poundmass (lbm) (*see* pound, avoirdupois)		
ppb (parts per billion)	(pure number)	$1* \cdot 10^{-9}$
ppm (parts per million)	(pure number)	$1* \cdot 10^{-6}$
ppt (parts per trillion)	(pure number)	$1* \cdot 10^{-12}$
psf (*see* pound-force per square foot)		
psi (*see* pound-force per square inch)		
psia (pound-force per square inch, absolute pressure) (*see* psi)		
psig (pound-force per square inch, gage pressure) (*see* psi)		
quad (quadrillion Btu$_{39°}$) [QBtu]	energy	1.059 67 EJ
quart (Imperial) (1/4 Imperial gallon) [qt]	volume	1.136 523 L
quart (U.S. dry) (1/32 U.S. bushel) [qt]	volume	1.101 221 L
quart (U.S. liquid) (1/4 U.S. gallon) [qt]	volume	0.946 352 9 L
R-value (°F·h·ft^2/Btu) [R-]	thermal insulance	0.176 K·m^2/W
rad (CGS) [rad, rd]	absorbed dose	0.01* Gy
radar nautical mile	time	12.261 ms
range (6 miles) (U.S. public land survey) [R]	length	9.6 km
ream (sheets of paper) [rm]	(pure number)	500*
rehoboam (champagne)	volume	4.5* L

Non-SI unit [Symbol]	Quantity	SI definition
rem (Roentgen equivalent man)	dose equivalent	0.01* Sv
revolution (360°) [r]	plane angle	2π rad
revolution per minute [r/min]	angular velocity	0.104 719 rad/s
rhe (CGS) (1/poise)	fluidity	10* 1/(Pa·s)
rod (U.S. survey) (16.5 U.S. survey feet) [rd]	length	5.029 210 m
roentgen (X-ray exposure) [R]		258 µC/kg
rood (1/4 acre)	area	1011.72 m²
rpm (see revolutions per minute)		
rydberg	energy	2.1799 aJ
salmanazar (champagne)	volume	9* L
scruple (apothecary) (20 grains) [sc]	mass	1.295 978 2* g
second (arc second) (1/60 arc minute) ["]	plane angle	4.848 137 µrad
second (right ascension) (1/86 400 circle) [ˢ]	plane angle	72.722 05 µrad
second (sidereal) [s]	time	≈0.997 269 6 s
section (1± mi²) (U.S. public land survey) [S, sec]	area	2.59± km²
shake	time	10* ns
shoe size (see size, shoe length)		
short ton (2000 pounds) [ton]	mass	907.184 74 kg
short ton of coal (consumption, EIA 1990)	energy	≈22.6 GJ
short ton of coal (production, EIA 1990)	energy	≈23.1 GJ
shot (1 fl oz U.S.)	volume	30 mL
size (shoe length, European)	insole length	6.67 mm·size
size (shoe length, U.S. children's)	insole length	9.8 mm · size + 102 mm
size (shoe length, U.S. men's)	foot length	8.5 mm · size + 186 mm
size (shoe length, U.S. women's)	foot length	8.5 mm · size + 178 mm
slug (lbf·s²/ft)	mass	14.593 903 kg
slug (MKpS) (kgf·s²/m) [hyl]	mass	9.806 65* kg
solar luminosities (stars)	power	≈382.7 YW
span (common) (9 inches)	length	23 cm
specific gravity (gases) (density relative to that of dry air at 20 °C) [sp gr]	density	1.204 kg/m³
specific gravity (solids and liquids) (density relative to that of water at 4 °C) [sp gr]	density	1000 kg/m³

Non-SI unit [Symbol]	Quantity	SI definition
speed of light [c]	speed	299.792 458* Mm/s
square (100 ft²)	area	9.3 m²
square chain (U.S. survey) [ch²]	area	404.687 261 m²
square degree [deg²]	solid angle	304.617 µsr
square foot (international) [ft²]	area	0.092 903 04* m²
square foot (U.S. survey) [ft²]	area	0.092 903 41* m²
square foot per hour [ft²/h]	diffusivity	25.8064* mm²/s
square foot per second [ft²/s]	kinematic viscosity	929.0304* m²/s
square inch (sq in) [in²]	area	6.4516* cm²
square mil [mil²]	area	645.16* µm²
square mile (sq mi) (international) [mi²]	area	2.589 988 km²
square mile (U.S. survey) [mi²]	area	2.589 998 km²
square rod (U.S. survey) [rod²]	area	25.292 85 m²
square yard (sq yd) [yd²]	area	0.836 127 4 m²
standard cubic foot (natural gas) [scf]	mass	≈20.2 g
statampere (CGS-esu)	current	333.564 pA
statcoulomb (CGS-esu)	charge	333.564 pC
statfarad (CGS-esu)	capacitance	1.112 650 pF
stathenry (CGS-esu)	inductance	898.7554 GH
statmho (CGS-esu)	conductance	1.112 650 pS
statohm (CGS-esu)	resistance	898.7554 GΩ
statvolt (CGS-esu)	potential	299.7924 V
stere (st) (obsolete metric name for cubic meter [m³])		
sthene (MTS) (t·m/s²) [sn]	force	1000* N
stilb (CGS) (cd/cm²) sb	luminance	10 000* cd/m²
stokes (CGS) (cm²/s) St	viscosity	1* cm²/s
stone (14 pounds) [st]	mass	6.350 kg
sub (paper) ("pound")	mass per area	3.75 g/m²
tablespoon (1/2 fl oz) [T, tbsp]	volume	15 mL
Tcf (trillion cubic feet)	volume	28.317 km³
teaspoon (1/6 fl oz) [t, tsp]	volume	5 mL
terawatt hour [TW·h]	energy	3.6* EJ
tex (thread)	mass per length	1* mg/m
therm (EU) (100 000 Btu_IT) [th]	energy	105.506 MJ
therm (U.S.) (100 000 Btu_{59°}) [th]	energy	105.4804* MJ
thou (mil) (0.001 inch)	length	25.4* µm
ton, assay [AT]	mass	29.167 g

Non-SI unit [Symbol]	Quantity	SI definition
ton, displacement (ships) [T]	mass	1016.047 kg
deadweight tons [DWT]	mass of cargo, fuel, and crew	
light displacement tons	mass of unloaded ship	
loaded displacement tons	DWT + light displacement tons	
ton (freight, shipping, or measurement) (40 ft³)	volume	1.13 m³
ton, long or gross (2240 lb) [T]	mass	1016.047 kg
ton, metric (tonne) [t]	mass	1000* kg
ton of energy (*see* metric ton of coal equivalent, short ton of coal, etc.)		
ton (nuclear weapons) (10⁹ cal$_{th}$) [t]	energy	4.184* GJ
ton, refrigeration (commercial) (12 000 Btu/h)	power	3517 W
ton, refrigeration (standard)	energy	303.9 MJ
ton, register (ships) (100 ft³)	volume	2.832 m³
gross register ton [GRT] ("gross weight")—total interior volume		
net register ton—volume of cargo space only		
ton, short or net (2000 lb)	mass	907.184 74 kg
ton (long) per cubic yard [T/yd³]	density	1.328 939 Mg/m³
ton (short) per cubic yard [ton/yd³]	density	1.186 553 Mg/m³
ton-force ("tonweight") (2000 lbf) [tonf]	force	8.896 443 kN
ton-force mile ("ton mile") [tonf·mi]	energy	14.32 MJ
tonne (metric ton) [t]	mass	1000* kg
torr (millimeter of mercury) (atm/760)	pressure	133.322 Pa
township (6 miles) (U.S. public land survey) [T]	length	9.6 km
township (36 sections) (U.S. public land survey) [twp]	area	93.2± km²
tun (252 U.S. gallons)	volume	0.954 m²

Non-SI unit [Symbol]	Quantity	SI definition
u (*see* atomic mass unit)		
unit pole	magnetic flux	125.6637 nWb
watt hour [W·h]	energy	3600* J
watt second [W·s]	energy	1* J
watt per square centimeter [W/cm²]	irradiance	10* kW/m²
watt per square inch [W/in²]	irradiance	1.550 003 kW/m²
week (7 d) [wk]	time	604.8* ks
X-unit (X rays)	length	100.2 fm
yard [yd]	length	0.9144* m
yd² (*see* square yard)		
yd³ (*see* cubic yard)		
year (365 d) [yr]	time	31.536 Ms
year (anomalistic or perihelion)	time	≈31.558 43 Ms
year (cosmic or galactic) (≈230 Ma)	time	≈7 Ps
year, leap (366 d) [yr]	time	31.6224 Ms
year (seasonal or tropical) (annum) (≈365.2422 d) [a]	time	≈31.556 926 Ms
year, sidereal (≈365.2563 d)	time	≈31.558 15 Ms

NONALPHABETIC UNIT SYMBOLS FOR NON-SI UNITS

'	*see* foot and arc second
"	*see* inch and arc minute
"Hg	*see* inch of mercury
"WC	*see* inch of water (column)
°	*see* degree
#	*see* pound (avoirdupois)
γ	gamma (obsolete symbol for microgram [μg])
γ	gamma (obsolete symbol for nanotesla [nT])
λ	lambda (obsolete symbol for microliter [μL] = cubic millimeter [mm³])
μ	obsolete symbol for micrometer (μm) (*see* micron)

ANSWERS TO PROBLEMS

Note: calculated values are rounded according to the rules in Chapter 3.

CHAPTER 1   INTRODUCTION

1. the International System of Units
2. Many metric units are not SI.
3. no conversions, no fractions, no long rows of zeroes, unambiguous letter symbols, coherent
4. coherent
5. non-SI units
6. inch-pound units
7. metric system (now SI)
8. in terms of SI base units
9. SI has only 1 unit for each quantity.

CHAPTER 2   SI STRUCTURE AND MATHEMATICAL CONVENTIONS

1. a physical attribute or phenomenon that can be measured
2. a standard of measurement
3. slanted (italic, or oblique)
4. upright (roman, or normal)
5. meter (m), length; kilogram (kg), mass; second (s), time; kelvin (K), temperature; ampere (A), electric current; mole (mol), amount of substance; candela (cd), luminous intensity
6. supplementary
7. derived
8. divided by
9. kg/s, kgs·s^{-1}, $\dfrac{kg}{s}$
10. kg·s
11. compound units
12. quantity
13. quantity
14. quantity
15. unit
16. unit
17. quantity
18. prefixes mega and larger, units named after a person, liter
19. D
20. C

21. D
22. D
23. D
24. C
25. A
26. B
27. B
28. A
29. C
30.

5		1	7	3	13
4	9	16		2	

31.

$5.98 \cdot 10^{24}$ kg		3		
	4	π	$(6.37 \cdot 10^{6}$ m$)^3$	

32. $2.6 \cdot 10^{-10}$

CHAPTER 3   PRECISION AND ROUNDING

1. significant
2. precision
3. exact
4. leading
5. ending zeroes to the left of the decimal point
6. accuracy
7. 3
8. 4
9. 2
10. 2 or 3
11. 4
12. 2, 3, 4, or 5
13. 2
14. 3
15. 3
16. 4
17. 1
18. 2, 3, or 4
19. 5
20. 4
21. 7

22. 23.7
23. 520
24. 458 000
25. 2 900 000
26. 4.178
27. 200

CHAPTER 4   PREFIXES

1. eliminate placeholding (nonsignificant) zeroes, easy to pronounce, unambiguous, simple symbols
2. prefix
3. scientific
4. terabull
5. petadog
6. megabuck
7. millipede
8. exaray
9. attoboy
10. picoboo
11. m (meter, milli), T (tera, tesla)
12. trillion
13. billion
14. a multiple (or submultiple) of the unit
15. 1 mm
16. 1 km
17. 1 μm
18. 1 ks
19. 1 MJ
20. 1 μm
21. 1 nm
22. 1 Mg
23. 1 Mm
24. 1 km
25. 1 mm
26. 8.7 km
27. 53 kW
28. 17 mW
29. 60 μs
30. 30 mg
31. 430 MJ
32. 700 GN
33. 6.3 ns
34. 2.35 m
35. 14.3 μPa
36. 1.8 s
37. 2.5 MJ

38. 2.3 pJ
39. 70 μm
40. 150 Gm
41. 35 K
42. 1.5 g
43. 7.9 GHz
44. 90 mA
45. 5 μV
46. 35 Mm
47. 591 Gm
48. 14 μW
49. 8.6 TW
50. 90 nm
51. 58 ps
52. 450 km
53. 62 mN
54. 3.8 ks
55. 260 kW
56. B
57. A

CHAPTER 5   LENGTH

1. 40 Mm
2. the distance light travels in a vacuum in 1/299 792 458 second
3. 300 Mm/s
4. 560 mm
5. 0.093 m
6. 1.6 cm
7. B
8. C
9. B
10. 100 000 times larger
11. 270 000 times farther
12. 150 mm (15 cm)
13. 73.4 km
14. 84.5 km
15. 4.40 m
16. 800 Pm

CHAPTER 6   AREA

1. area
2. hectare (ha)
3. 10^4 m^2
4. 10^{-4} m^2
5. 10^{-6} m^2
6. 10^4 m^2

7. 1 m^2

8. 1 km^2

9. $4.5 \text{ ha} = 4.5 \text{ hm}^2$

10. 79 cm^2

11. 9 cm^2

12. 71 km^2

13. 8 km^2

14. 6.4 Mm^2

15. $32 \text{ ha} = 32 \text{ hm}^2$

16. 80 m^2

17. 59 mm^2

18. $18 \text{ ha} = 18 \text{ hm}^2$

19. $7.1 \text{ ha} = 7.1 \text{ hm}^2$

20. 35 cm^2

21. 4500 m^2

22. 2.0 m^2

23. 41 cm^2

24. $4 \text{ ha} = 4 \text{ hm}^2$

25. 9.84 m^2

26. $6.6 \text{ ha} = 6.6 \text{ hm}^2$

27. 1800 cm^2

28. 510 Mm^2

CHAPTER 7   VOLUME

1. volume

2. cubic meter

3. liter

4. milliliter

5. microliter

6. cm^3

7. dm^3

8. mm^3

9. B

10. 10^{-6} m^3

11. 10^6 m^3

12. 10^3 m^3

13. 10^{-9} m^3

14. 10^{-3} m^3

15. 10^{-3} m^3

16. 10^{-6} m^3

17. 10^{-9} m^3

18. D

19. C

20. C

21. B

22. B

23. C

24. A

25. $1 \text{ L} = 1 \text{ dm}^3$

26. 1 dam^3

27. 1 hm^3

28. $1 \text{ L} = 1 \text{ dm}^3$

29. 1 m^3

30. $1 \text{ mL} = 1 \text{ cm}^3$

31. 53 m^3

32. $20 \text{ mL} = 20 \text{ cm}^3$

33. $60 \text{ mL} = 60 \text{ cm}^3$

34. $25 \text{ μL} = 25 \text{ mm}^3$

35. $40 \text{ L} = 40 \text{ dm}^3$

36. 68 km^3

37. $83 \text{ mm}^3 = 83 \text{ μL}$

38. 71 km^3

39. $36 \text{ cm}^3 = 36 \text{ mL}$

40. $4.7 \text{ L} = 4.7 \text{ dm}^3$

41. 120 m^3

42. $9 \text{ dam}^3 = 9000 \text{ m}^3$

43. 1.2 hm^3

44. $630 \text{ cm}^3 = 630 \text{ mL}$

45. $1.37 \text{ L} = 1.37 \text{ dm}^3$

46. 260 km^3

47. $6.1 \text{ L} = 6.1 \text{ dm}^3$

48. 2.6 hm^3

49. $7.2 \text{ L} = 7.2 \text{ dm}^3$

50. $140 \text{ cm}^3 = 140 \text{ mL}$

51. $22 \cdot 10^{18} \text{ m}^3 = 22 \text{ Mm}^3$

52. $1.4 \cdot 10^{27} \text{ m}^3 = 1.4 \text{ Gm}^3$

53. 32 hm^3

54. $5.3 \cdot 10^{15} \text{ m}^3$

55. $9 \cdot 10^{-31} \text{ m}^3$

56. $5.7 \text{ L} = 5.7 \text{ dm}^3$

57. $3.79 \text{ L} = 3.79 \text{ dm}^3$

58. 2.0 hm^3

59. 16 km^3

CHAPTER 8   MASS AND DENSITY

1. mass

2. metric ton, tonne

3. the mass of a liter of water at 4 °C

4. the mass of the International Prototype Kilogram

5. 8 Mg

6. 5 kg

7. 1.5 g

8. 54 g

9. 15 Mg

10. 20 μg

11. 70 mg

12. 86 Gg

13. 34 Tg

14. 6.3 g

15. 4.5 Mg

16. 13 mg

17. 6.5 kg

18. 12 Mg

19. 30 g

20. 7 g

21. 8.9 mg

22. 29 kg

23. 470 Tg

24. 31 mg

25. 62 Pg

26. 44 Gg

27. mass per volume

28. C

29. 1.3 kg/L

30. 3.2 kg/m^3

31. 4.78 kg/L = 4780 kg/m^3

32. 1.4 kg/L = 1.4 Mg/m^3

33. 2.6 kg/L = 2.6 Mg/m^3

34. 0.6 kg/L = 600 kg/m^3

35. 5.8 kg/L = 5.8 Mg/m^3

36. 5.5 kg/L = 5.5 Mg/m^3

37. 0.6 kg/L = 600 kg/m^3

38. float

39. 1.7 g/L = 1.7 kg/m^3

40. 2.8 ng/L = 2.8 μg/m^3

41. 10 Mg/m^3

42. 31 kg

43. 280 g

44. 41 Tg

45. 46.7 cm^3 = 46.7 mL

46. 17 cm^3 = 17 mL

47. 13 L

48. 35 mg/kg

49. 30 g/kg

50. 17 μg/kg

51. 2 g/L = 2 kg/m^3

52. 12 mg/L = 12 g/m^3

53. 55 ng/L = 55 μg/m^3

54. 1.8 kg

55. 2.0 kg

56. 460 Gg

57. 16 Mg

58. F

CHAPTER 9   TIME AND RATES

1. period

2. second

3. E

4. annum (a)

5. 1/86 400 mean solar day in 1900

6. Earth's rotation is slowing due to tidal friction.

7. 10 m/s

8. 5.7 ks = 1.6 h

9. 1.28 s

10. 130 Ms = 4.2 a (years)

11. 6.3 · 10^{12} m^3 = 6300 km^3

12. 6 ks = 1.7 h

13. 0.2 m/s = 200 mm/s

14. 16.9 ks

15. 28.2 ks

16. 1.66 Ms

17. 110 Ms

18. 7.9 Ps

19. 25 m/s

20. 25 m/s

21. 2.2 L/s

22. 64 L/s

23. 3000 m^3/s = 3 dam^3/s

24. 2.7553 h

25. 3 h 41 min 31 s

CHAPTER 10   FORCE

1. force

2. acceleration

3. m/s^2

4. $F = ma$

5. N = kg·m/s^2

6. force of gravity

7. 9.8 m/s^2

8. 380 N

9. 9.8 kN

10. 740 N

11. 80 kg

12. 12 kN
13. 120 N
14. 55 kg
15. 16 kg
16. 2.61 m/s^2
17. electromagnetic
18. gravitation
19. strong
20. 35 ZN
21. 740 N
22. 50 N·m
23. 689 N
24. 350 N

CHAPTER 11 ENERGY

1. energy
2. $E = Fd$
3. J = N·m
4. 2 J
5. 2 kJ
6. 20 MJ
7. 25 GJ
8. 500 kJ
9. 100 J
10. 15 m/s
11. 383 YJ
12. 630 kJ
13. 630 J
14. 1.2 GJ
15. 1.2 GJ
16. 1.3 MJ
17. 130 TJ
18. 340 J
19. 4 GJ
20. 4.0 GJ

CHAPTER 12 POWER

1. $P = E/t$
2. W = J/s
3. 120 kW
4. 42 kW
5. 360 kJ
6. 140 W
7. 2.1 TW
8. 32 kW
9. 180 W
10. 1.3 Gs = 40 a (years)

11. 51 MW
12. 58%
13. 25 kW
14. 100 kW
15. 100 W
16. 2.9 TW
17. 1.2 kW

CHAPTER 13 PRESSURE AND STRESS

1. $p = F/A$
2. Pa = N/m^2
3. 101 kPa
4. 450 Pa
5. 3 kPa
6. 70 kPa
7. 1.6 kN
8. 350 kPa
9. 52 EN
10. 98.5 kPa
11. 500 kPa
12. 20 kPa
13. 240 kPa
14. 20 kPa
15. 44 Mpa

CHAPTER 14 TEMPERATURE AND HEAT

1. heat
2. temperature
3. kelvin
4. degrees Celsius
5. joule
6. absolute zero
7. no negative numbers
8. 300 K
9. 100 K
10. 50 °C
11. 15 MK
12. 0 K = –273 °C
13. 273 K = 0 °C
14. 373 K = 100 °C
15. 293 K = 20 °C
16. 310 K = 37 °C
17. 6 kK = 6000 °C
18. 331 K = 58 °C
19. 185 K = –88 °C
20. 313 K = 40 °C

21. C

22. 400 W

23. 1.8 kW

24. 11 GJ

25. 690 TJ

26. triple point

27. the temperature of the triple point of water

28. 350 K

29. 4.87 zJ

30. 500 m/s

CHAPTER 15   FREQUENCY

1. frequency

2. $f = 1/t$

3. $Hz = 1/s$

4. $v = \lambda f$

5. $c = \lambda f$

6. 20 Hz to 20 kHz

7. electromagnetic radiation

8. 67 mHz

9. 15 MHz

10. 4.2 m/s

11. 1.3 m

12. 12 kHz

13. 400 THz

14. 5.5 pm

15. 1.5 GHz

16. infrared

17. blue

18. 3 fJ

19. 320 zJ

CHAPTER 16   ANGLES

1. radian

2. degree

3. grad

4. steradian

5. 2π

6. 4π

7. 4.07 rad

8. 163°

9. 68°

10. 35.278°

11. 18°20′52″

12. 2.170 12 rad

13. $12 \cdot 10^{-6}$

14. 35 nmi

15. 333 km

16. 6.2 m/s

17. 0.6

18. 7.4°

19. 30 mrad

20. 14°

21. 44 mrad

CHAPTER 17   ELECTROMAGNETIC UNITS

1. charge

2. current

3. voltage or potential

4. 160 zC

5. A

6. 570 mN

7. ampere (A)

8. $9.4 \cdot 10^{19}$

9. volt (V)

10. 1.8 W

11. $V = IR$

12. ohm (Ω)

13. conductors

14. insulators

15. 12 Ω

16. 2 A

17. 240 V

18. 120 V

19. 50 V

20. 100 mA

CHAPTER 18   THE MOLE

1. $6.02 \cdot 10^{23}$

2. $6.02 \cdot 10^{23}$

3. $6.02 \cdot 10^{23}$

4. $6.02 \cdot 10^{20}$

5. $6.02 \cdot 10^{17}$

6. $6.02 \cdot 10^{26}$

7. 1/12 the mass of a carbon-12 atom

8. 1.66 yg

9. 1 u

10. 1 u

11. 1 u

12. mol = g/u

13. 20 yg

14. 16 g

15. 58 g

16. 18 g
17. 54 g
18. 1.4 Mg
19. 5 mol
20. $1.2 \cdot 10^{24}$

CHAPTER 19   PHYSIOLOGICAL UNITS

1. 540 THz
2. lumen (lm)
3. candela (cd)
4. candle, candlepower
5. 1000 lm
6. 135 cd
7. 250 lx
8. 50 lm
9. 490 μg
10. 23 mGy
11. 740 mBq/L
12. 400 μSv
13. 3 mGy

BIBLIOGRAPHY

American National Standards Institute and Institute of Electrical and Electronics Engineers, Inc. (1992), *American National Standard for Metric Practice*, ANSI/IEEE std 268-1992.

International Organization for Standardization (1992), *Quantities and Units*, ISO publication 31 (14 volumes).

National Institute of Standards and Technology (1991), *The International System of Units (SI)*, NIST Special Publication 330-1991.

National Institute of Standards and Technology (1992), *Interpretation of the SI for the United States and Metric Conversion Policy for Federal Agencies*, NIST Special Publication 814-1992.

U.S. Metric Association (1993), *Guide to the Use of the Metric System (SI Version)*, USMA, Northridge, California.

ABOUT THE AUTHOR

Dennis Brownridge holds a Ph.D. in geography from the University of Oregon and formerly taught at the University of California at Santa Barbara. He now devotes his time to writing on both technical and popular subjects and teaches at a private school in Arizona. His work has given him experience with a great variety of traditional units whose impracticality led him to appreciate the simplicity of the International System. He has taught SI for many years and has given seminars on the topic.

Index

For non-SI units, see also Appendix B.

Abbreviations, 78
Absolute pressure, 49
Absolute zero, 52, 56
Absorbed dose, 73
Acceleration, 38
Acceleration, angular, 75
Accuracy, 9
Alternative names, 25, 30
Ampere, 65
Angles, 62
Annum, 35
Arc minute, 62
Arc second, 62
Area, 21
Atan, 63
Atmosphere, 49
Atomic mass unit, 69
Atoms, 69
Avogadro's number, 69

Babylonian units, 2, 62
Base units, 4
Becquerel, 73
Bending moment, 75
Billion, 11
BIPM, 30
Brightness, 72
Buoyancy, 32

Calculators, 6, 14, 35
Candela, 72
Capacitance, 67
Capitalization, 77
Celsius, 52
CGPM, 1, 56
Charge, 65
Coefficient of heat transfer, 75
Coherent system, 1
Compound units, 5, 75, 79
Concentration, 32, 75
Conductance, 67
Conductivity, 75
Conductor, 66
Conversion factors, 80
Coulomb, 65
Coulomb's Law, 65
Counting numbers, 9
Curie, 73
Current, 65
Current density, 75

Day, 35
Deci, 25
Decimal places, 8
Decimal points, 13, 78
Degree, 62
Degree Celsius, 52
Deka, 25
Density, 31, 75
Derived units, 4
Digit separators, 5, 78
Dimensionless units, 62
Discharge, 75
Division, 79
Dose equivalent, 73

E (calculator key), 6, 14
Earth, 17, 35, 38, 46
Efficiency, 47
Einstein's equation, 43
Electrical energy, 43
Electricity, 65
Electromagnetic force, 40
Electromagnetic radiation, 58, 72
Electromagnetic units, 65
Electromotive force, 66
Electrons, 65
Electroweak force, 40
Energy, 43, 46, 75
Energy states, 55
ENG display, 11
English system, 2
Entropy, 75
Exact numbers, 8
Exponents, 5, 11
Exposure, 75

Factor label method, 80
Farad, 67
Flow, 36, 75
Food, 43
Force, 38, 75
Force of gravity, 30, 38
Fraction bars, 5
Fractions, 79
Frequency, 58
Fuel consumption, 75

g, 38
Gage pressure, 49
Gamma rays, 43, 59

Gases, 55
Gigannum, 35
GMT, 35
Grad, 63
Gradient, 63
Gravitation, 35, 38, 40, 65
Gray, 73
Greenwich time, 35

Heat, 43, 52, 75
Heat capacity, 75
Heat flow, 54
Heat flux, 46
Hectare, 21
Henry, 67
Hertz, 58
Hour (time), 35

Illuminance, 72
Illumination, 72
Impulse, 76
Inch-pound units, 2
Inductance, 67
Infrared, 43, 59
Insolation, 76
Insulation, 54
Insulator, 66
Intensity, 54
International System, 1
Inverse tangent, 63
Ionizing radiation, 72
IPTS, 56
Irradiance, 54, 76
Irradiation, 76
Italics, 4, 5, 77

Joule, 43, 46, 52

Kelvin, 52, 56
Kiloannum, 35
Kilogram, 30
Kilometer, 17
Kinetic energy, 44, 52, 55
Knot, 63

Latent heat, 55
Latitude, 63
Length, 17
Letter style, 77
Light, 43, 59, 72

Liter, 25
Litre, 77
Lumen, 72
Luminance, 76
Luminosity, 46
Luminous flux, 72
Luminous intensity, 72
Lux, 72

Magnetic field strength, 76
Magnetic flux, 67
Mass, 30, 38, 76
Mass-energy equivalence, 43
Matter, 30
Mean solar time, 35
Megannum, 35
Metabolism, 46
Meter, 17
Metre, 77
Metric Conversion Act, 2
Metric system, 1
Metric ton, 30
Microwaves, 59
Milliliter, 25
Minute (angle), 62
Minute (time), 35
Mixtures, 32
Molar, 75
Mole, 69
Molecule, 55
Moment, 76
Momentum, 76
Month, 35
Multiples of units, 13
Multiplication, 79

Nautical mile, 63
Neutrons, 69
Newton, 38, 49
Non-SI units, 1, 81
Notations, 11
Nuclear energy, 43
Nuclear force, 40
Nucleons, 69

Ohm, 66
Ohm's Law, 66
Omnibus Trade Act, 2
Ordinary notation, 11

Parts per million, 32
Pascal, 49

Per, 4
Percent, 32
Permeability, 76
Permittivity, 76
Photon, 60
Physiological units, 72
Planck's Law, 60
Plurals, 77
Potential, 66
Power, 46, 76
Powers of ten, 11, 78
Precision, 8
Prefixes, 11, 17, 21, 25, 26, 30,
    32, 78, 79
Pronunciation, 77
Protons, 65, 69
Punctuation, 77

Quantity, 4

Rad, 73
Radian, 62
Radiance, 76
Radiant energy, 72
Radiation, 40, 43, 72
Radio waves, 43, 59
Radioactivity, 73
Rate, 75
Relative mass, 69
Reluctance, 76
Rem, 73
Resistance, 66
Resistivity, 76
Rounding, 8

Scientific notation, 11
Second, 35
Second (angle), 62
Second Law of Motion, 38, 43
SI, 1
Siemens, 67
Sievert, 73
Significant digits, 8
Slope, 63
Sound, 43, 58
Spacing, 5, 77
Specific, 75
Specific heat, 55
Speed, 4, 36
Speed of light, 17, 43
Spelling, 77

Square meter, 21
Steradian, 62
Stress, 49
Strong force, 40
Substance, 69
Supplementary units, 4, 62
Surface tension, 76
Symbols, 4, 12, 77, 78

TAI, 35
Temperature, 52
Tesla, 67
Thermal, 75
Thermal conductance, 54
Thermal insulance, 54
Thrust, 76
Time, 35, 58
Tonne, 30
Torque, 40, 43, 76
Transmissivity, 76
Trillion, 11
Triple point, 56

Ultraviolet, 43, 59
Unit, 4
UTC, 35

Vacuum, 49
Velocity, 36, 38, 76
Viscosity, 76
Volt, 66
Voltage, 66
Volume, 25, 31

Water, 56
Watt, 46
Wave law, 58
Wave number, 76
Wavelength, 58, 72
Weak force, 40
Weber, 67
Weight, 30, 38
Work, 43

X rays, 43, 59

Year, 35

Zeroes, 4, 8, 78
Zulu time, 35